More
Picture-Perfect
SCIENCE
Lessons

More
Picture-Perfect
SCIENCE
Lessons

Using Children's Books  to Guide Inquiry, K–4

By Karen Ansberry
Emily Morgan

NSTApress

NATIONAL SCIENCE TEACHERS ASSOCIATION

Arlington, Virginia

NATIONAL SCIENCE TEACHERS ASSOCIATION

Claire Reinburg, Director
Judy Cusick, Senior Editor
Andrew Cocke, Associate Editor
Betty Smith, Associate Editor
Robin Allan, Book Acquisitions Coordinator

ART AND DESIGN
Will Thomas, Director
Cover, Inside Design, and Illustrations by Linda Olliver
Photographs by Karen Ansberry and Emily Morgan

PRINTING AND PRODUCTION
Catherine Lorrain, Director
Nguyet Tran, Assistant Production Manager
Jack Parker, Electronic Prepress Technician

NATIONAL SCIENCE TEACHERS ASSOCIATION
Gerald F. Wheeler, Executive Director
David Beacom, Publisher

Library of Congress Cataloging-in-Publication Data

Ansberry, Karen Rohrich, 1966-
  More picture-perfect science lessons : using children's books to guide inquiry, Grades K-4 / by Karen Ansberry and Emily Morgan.
    p. cm.
  Includes index.
  ISBN-13: 978-1-933531-12-0
  ISBN-10: 1-933531-12-6
  1. Science--Study and teaching (Elementary) 2. Picture books for children--Educational aspects. I. Morgan, Emily R. (Emily Rachel),
1973- II. Title.
LB1585.A57 2007
372.3'5--dc22
                              2007007200

Contents

Foreword

In the mid-1980s, BSCS developed the 5E instructional model. At that time we adapted and extended a model from the Science Curriculum Improvement Study (SCIS) which was a contemporary science curriculum for elementary schools. The 5E model became a standard feature of BSCS programs, beginning with the program for which it was originally designed. That elementary program is now known as *BSCS Science Tracks: Connecting Science and Literacy.*

At the time we developed the BSCS 5E model, we had no idea about National Science Education Standards, the reemergence of inquiry, No Child Left Behind (NCLB) with its emphasis on reading literacy, and the dominating influence that assessment would have on science education. But, the decade of the 1990s did present changes and challenges for science education, especially programs in elementary schools. With NCLB we have witnessed more and more emphasis on reading in lower elementary grades with the result of less and less science. Of course, concerns mounted in the science education community. Statements about the worth and importance of science were heard. "Elementary should realize the importance of science." "Science can enhance literacy." But, the plea went unheard. Or, they were heard and acknowledged without any change in science instruction. Why was this? Teachers provided an answer to the question when they asked—"Where are the lessons?" The teaching community needed examples, in the form of curriculum materials, of how to incorporate literacy into science instruction. This book presents one response to the elementary teacher's question.

Karen Ansberry and Emily Morgan present the science education community with a refreshing and positive remedy to the reduction of science teaching in elementary schools. In *More Picture Perfect Science Lessons: Using Children's Books To Guide Inquiry, K–4,* they present an integrated instructional approach that addresses National Science Education Standards, inquiry, and the need for elementary teachers to enhance the reading skills of children. In short, the authors use the BSCS 5E instructional model to present science lessons. In doing so, they integrate reading strategies. The activities complement topics included in most school science programs—rocks, trees, magnets, and plants.

The authors have contributed to the goal of more and better science instruction in elementary schools in the United States. This book presents lessons that accommodate every elementary teacher's need to be efficient. You can teach so children learn science AND develop reading abilities.

Rodger W. Bybee
Executive Director, BSCS
Colorado Springs, Colorado

Preface

A class of third-grade students laughs as their teacher reads Doreen Cronin's *Diary of a Worm*. Students are listening to the earthworm reading his diary, "June 15th: My older sister thinks she's so pretty. I told her that no matter how much time she spends looking in a mirror, her face will always look just like her rear end." The third-grade class giggles as the teacher continues to read the worm's hilarious diary entries. After the read aloud, the teacher leads students through a reading comprehension strategy called *questioning the author* (Beck et al. 1997) in which the students learn to think critically about what they are reading. The teacher models this by generating a list of questions to ask the author, such as "Is this accurate—a worm's head and tail look just alike? Can you tell a worm's head from its tail?" Students then observe live earthworms with hand lenses and read a nonfiction book about worms in an effort to find the answer. Through this exciting lesson, students construct their own understandings about earthworm adaptations, how earthworms help the earth, and how to design and conduct simple experiments to answer questions.

What Is Picture-Perfect Science?

This scenario describes how a children's picture book can help guide students through an engaging, hands-on inquiry lesson. *More Picture-Perfect Science Lessons* contains 15 science lessons for students in kindergarten through grade four, with embedded reading comprehension strategies to help them learn to read and read to learn while engaged in inquiry-based science. To help you teach according to the National Science Education Standards, the lessons are written in an easy-to-follow format for teaching inquiry-based science: the Biological Sciences Curriculum Study 5E Instructional Model (Bybee 1997). This learning-cycle model allows students to construct their own understandings of scientific concepts as they cycle through the following phases: *engage, explore, explain, elaborate,* and *evaluate. More Picture-Perfect Science Lessons* is primarily a book for teaching science, but reading-comprehension strategies are embedded in each lesson. You can model these essential strategies throughout while you keep the focus of the lessons on science.

Use This Book Within Your Science Curriculum

We wrote *More Picture-Perfect Science Lessons* to supplement, not replace, your school's existing science program. Although each lesson stands alone as a carefully planned learning cycle based on clearly defined science objectives, the lessons are intended to be integrated into a complete curriculum in which concepts can be more fully developed. The lessons are not designed to be taught sequentially. We want you to use *More Picture-Perfect Science Lessons* where appropriate within your school's current science program to support, enrich, and extend it. And we want you to adapt the lessons to fit your school's curriculum, the needs of your students, and your own teaching style.

Special Features of the Book

1 Ready-To-Use Lessons With Assessments

Each lesson contains background for the teacher, engagement activities, hands-on explorations, student pages, suggestions for student and teacher explanations, opportunities for elaboration, assessment suggestions, and annotated bibliographies of more books to read on the topic. Assessments range from poster sessions with rubrics to student-created books to formal multiple-choice and extended-response quizzes.

2 Reading Comprehension Strategies

Reading comprehension strategies based on the book *Strategies that Work* (Harvey and Goudvis 2000) and specific activities to enhance comprehension are embedded throughout the lessons and clearly marked with an icon (   ). Chapter 2 describes how to model these strategies while reading aloud to students.

3 Standards-Based Objectives

All lesson objectives were adapted from *National Science Education Standards* (NRC 1996) and are clearly identified at the beginning of each lesson. Chapter 5 outlines the National Science Education Standards for K–4 and shows the correlation between the lessons and the Standards.

4 Science as Inquiry

As we said, the lessons in *More Picture-Perfect Science Lessons* are structured as guided inquiries following the 5E model. Guiding questions are embedded throughout each lesson and marked with an icon (?). The questioning process is the cornerstone of good teaching. A teacher who asks thoughtful questions arouses students' curiosity, promotes critical thinking skills, creates links between ideas, provides challenges, gets immediate feedback on student learning, and helps guide students through the inquiry process. Each lesson includes an Inquiry Place, a section at the end of the lesson that suggests ideas for developing open

inquiries. Chapters 3 and 4 explore science as inquiry and the BSCS 5E Instructional Model.

References

Beck, I., R. Hamilton, L. Kucan, and M. McKeown. 1997. *Questioning the author: An approach for enhancing student engagement with text.* Newark, DE: International Reading Association.

Bybee, R. W. 1997. *Achieving scientific literacy: From purposes to practices.* Portsmouth, NH: Heinemann.

Harvey, S. and A. Goudvis. 2000. *Strategies that work: Teaching comprehension to enhance understanding.* York, ME: Stenhouse Publishers.

National Research Council. 1996. *National Science Education Standards.* Washington, DC: National Academy Press.

Children's Book Cited

Cronin, D. 2003. *Diary of a worm.* New York: Joanna Cotler Books.

Editors' Note:

More Picture-Perfect Science Lessons builds upon the texts of 29 children's picture books to teach science. Some of these books feature animals that have been anthropomorphized—forest animals talk, a worm keeps a diary. While we recognize that many scientists and educators believe that personification, teleology, animism, and anthropomorphism promote misconceptions among young children, others believe that removing these elements would leave children's literature severely underpopulated. Further, backers of these techniques not only see little harm in their use but also argue that they facilitate learning.

Because *More Picture-Perfect Science Lessons* specifically and carefully supports scientific inquiry—"That Magnetic Dog" lesson, for instance, teaches students how to weed out misconceptions by asking them to point out inaccurate statements about magnetism—we, like our authors, feel the question remains open.

Acknowledgments

We would like to give special thanks to science consultant Carol Collins for sharing her expertise in teaching inquiry-based science, for giving us many wonderful opportunities to share Picture-Perfect Science with teachers, and for supporting and encouraging our efforts.

We would also like to express our gratitude to language arts consultant Susan Livingston for opening our eyes to the power of modeling reading strategies in the content areas and for teaching us that every teacher is a reading teacher.

We deeply appreciate the care and attention to detail given to this project by Claire Reinburg, Betty Smith, and Catherine Lorrain at NSTA Press.

Many thanks to Linda Olliver for her fantastic illustrations and cover art.

And these thank yous as well:

- To Nancy Landes for helping us to better understand the BSCS 5E Model.

- To all of the wonderful teachers who field-tested these lessons and gave us feedback for improvement, including Janell Bishop, Michelle Byrd, Megan Calpin, Emily Chandler, Patti Custis, Tammy DiPenti, Anna Flaig, Michelle Gallite, Shirley Hudspeth, Jen Molitor, Patty Quill, Kristin Riekels, Jill Requejo, Jodie Scalfaro, Diana Suit, and Karen Vome.

- To the teachers at Mason Early Childhood Center, Mason Heights Elementary, Western Row Elementary, and Mason Intermediate School in Mason City School District for inspiring our efforts to integrate literature and science.

- To the teachers at Elda Elementary and Morgan Elementary in Ross Local Schools, Donna Shartzer and the teachers at Breckenridge County Schools, and Sharon Watkins and the teachers at Norwalk City Schools for enthusiastically participating in our workshops.

- To Carol Ogden at Mason City Schools for giving us the inspiration for the UV beads inquiry and for promoting our books and workshops to everyone she meets!

- To Lois Cropenbaker at Ross Local Schools for being our "#1 Picture-Perfect Fan."

- To Nancy Borchers at Ross Local Schools for providing us with opportunities to share our program with teachers and for her continued support.

- To Kendall Hauer at Miami University and Linda Sutphin at Mason City Schools for helping to make our geology lesson "rock solid."

- To Bill Robertson for sharing his expertise in physical science.

- To Don Kaufmann, Lisa Streit, and Mike Wright at Miami University's GREEN Teachers Institute for supporting our workshops and to the GREEN Institute participants for spending two dynamic weeks with us.

- To Mary Flower for her legal advice.

- To Claire Ansberry, Noreen Bouley, Jenny and Tom Doerflein, Randy Doughman, Anna

Flaig, Principal Dan Hamilton at Sherwood Elementary, Megan Kessler, Michele Klysz-Robben, Amy Moon, Yvonne Slusser, Karen Thompson, and Marylee and Jim Vennemeyer for "kid-wrangling" at our lively after-school photo shoots.

- To Jim Stevens, Sherry Stoffer, Diana Suit, and Jim Vennemeyer for contributing some of the photographs used in the book.

- To Rhonda Vanderbeek for help with photo editing.

- To Sherry Stoffer for designing our logo.

- To Christopher and Jeanette Canyon for enthusiastically sharing their talents with our students and workshop participants and for their continued advice and encouragement.

- To April Pulley Sayre for sharing her sense of wonder and inspiring us with her eloquent books.

- To Vicki Cobb for sharing her passion for science with our workshop participants.

- To Loren Long, whose work continues to inspire us.

- To Linda Keller at Barnes & Noble in West Chester, Ohio, for providing us with opportunities to share our books with teachers.

- To John Hutton and Sandra Gross at Blue Manatee Children's Bookstore in Cincinnati for their support and for sharing *Last Child in the Woods* with us.

- To Julie Valin at Dawn Publications, Jodee Seibert at Heinemann Library, and Maranda McCarthy at Picture Window Books for supplying us with high-quality science-related picture books to preview.

- To Jackie Collier at the University of Dayton for teaching her students that all teachers are reading teachers.

- To Gene Easter for introducing us to *That Magnetic Dog* and "The Day My Feet Were Magnets".

- To Ken Roy, Director of Environmental Health and Safety, Glastonbury (Connecticut) Public Schools, and Mark Dykewicz, M.D., at St. Louis University School of Medicine, for their advice on safety issues.

- To our families and friends for their moral support.

- To our husbands, Kevin Ansberry and Jeff Morgan, for their patience and encouragement.

- To our grandmothers for always believing in us.

- And to our parents, who were our very first teachers.

The contributions of the following reviewers are also gratefully acknowledged: Mariam Jean Dreher, DeLene Hoffner, Rich Hogen, Bill Robertson, and Christine Anne Royce.

About the Authors

Karen Ansberry is the Elementary Science Curriculum Leader and a former fifth- and sixth-grade science teacher at Mason City Schools, in Mason, Ohio. She has a Bachelor of Science in Biology from Xavier University and a Master of Arts in Teaching from Miami University. Karen lives in historic Lebanon, Ohio, with her husband, Kevin, and their two dogs and two cats.

Emily Morgan is a science consultant for the Hamilton County Educational Service Center. She formerly taught second- through fourth-grade science lab at Mason City Schools in Mason, Ohio, and seventh-grade science at Northridge Local Schools in Dayton, Ohio. She has a Bachelor of Science in Elementary Education from Wright State University and a Master of Science in Education from the University of Dayton. Emily lives in West Chester, Ohio, with her husband, Jeff, and their dog and two cats.

Karen Ansberry and Emily Morgan are the authors of *Picture-Perfect Science Lessons: Using Children's Books to Guide Inquiry (Grades 3-6)* published by NSTA Press in 2005. In collaboration with language arts consultant Susan Livingston, they received a Toyota Tapestry Award for their *Picture-Perfect Science* grant proposal in 2002.

Emily and Karen share a passion for science, nature, animals, travel, teaching, and children's literature. They enjoy working together to facilitate Picture-Perfect Science teacher workshops. This is their second book.

For more information on Picture-Perfect Science teacher workshops, go to:

www.pictureperfectscience.com

About the Picture-Perfect Science Program

The Picture-Perfect Science program originated from Emily Morgan's and Karen Ansberry's shared interest in using children's literature to make science more engaging. In her 2001 master's thesis study involving 350 of Emily's third grade science lab students at Western Row Elementary, she found that students who used science trade books instead of the textbook scored significantly higher on district science performance assessments than students who used the textbook only. Convinced of the benefits of using picture books to engage students in science inquiry and to increase science understanding, Karen and Emily began collaborating with Sue Livingston, Mason's elementary language arts curriculum leader, in an effort to integrate literacy strategies into inquiry-based science lessons. They received grants from the Ohio Department of Education (2001) and Toyota Tapestry (2002) in order to train all third grade through sixth grade science teachers, and in 2003 also trained seventh and eighth grade science teachers with district support. The program has been presented both locally and nationally, including at the National Science Teachers Association national conferences in San Diego, Philadelphia, Dallas, and Nashville.

For more information on Picture-Perfect Science teacher workshops, go to: *www.pictureperfect-science.com*

Why Read Picture Books in Science Class?

Think about a book you loved as a child. Maybe you remember the zany characters and rhyming text of Dr. Seuss classics like *One Fish Two Fish Red Fish Blue Fish* or the clever poems in Shel Silverstein's *Where the Sidewalk Ends*. Perhaps you enjoyed the page-turning suspense of *The Monster at the End of This Book* or the fascinating facts found in Aliki's *Digging Up Dinosaurs*. You may have seen a little of yourself in *Where the Wild Things Are, Ramona the Pest,* or *Curious George.* Maybe your imagination was stirred by the colorful illustrations in *The Very Hungry Caterpillar* or the stunning photographs in Seymour Simon's *The Moon.* You probably remember the warm, cozy feeling of having a treasured book like *Frog and Toad Are Friends* or *Charlotte's Web* being read to you by a parent or grandparent. But chances are your favorite book as a child was *not* your third-grade science textbook. The format of picture books offers certain unique advantages over textbooks and chapter books for engaging students in a science lesson. More often than other books, fiction and nonfiction picture books stimulate students on both the emotional and intellectual levels. They are appealing and memorable because children readily connect with the imaginative illustrations, vivid photographs, experiences and adventures of

Teachers enjoy using picture books.

characters, engaging storylines, the fascinating information that supports them in their quest for knowledge, and the warm emotions that surround the reading experience.

What characterizes a picture book? We like what *Beginning Reading and Writing* says, "Picture books are unique to children's literature as they are defined by format rather than content. That is, they are books in which the illustrations are of equal importance as or more important than the text in the creation of meaning" (Strickland and Morrow 2000, p. 137). Because picture books are more likely to hold children's attention, they lend themselves

to reading comprehension strategy instruction and to engaging students within an inquiry-based cycle of science instruction. "Picture books, both fiction and nonfiction, are more likely to hold our attention and engage us than reading dry, formulaic text. … engagement leads to remembering what is read, acquiring knowledge and enhancing understanding" (Harvey and Goudvis 2000, p. 46). We wrote *More Picture-Perfect Science Lessons* (and the first volume, *Picture-Perfect Science Lessons*) so teachers can take advantage of the positive features of children's picture books by supplementing the traditional science textbook with a wide variety of high-quality fiction and nonfiction science-related picture books.

The Research

1 Context for Concepts

Literature gives students a context for the concepts they are exploring in the science classroom. Children's picture books, a branch of literature, have interesting storylines that can help students understand and remember concepts better than they would by using textbooks alone, which tend to present science as lists of facts to be memorized (Butzow and Butzow 2000). In addition, the colorful pictures and graphics in picture books are superior to many texts for explaining abstract ideas (Kralina 1993). As more and more content is packed into the school day and higher expectations are placed on student performance, it is critical for teachers to teach more in the same amount of time. Integrating curriculum can help accomplish this. The wide array of high-quality children's literature available can help you model reading comprehension strategies while teaching science content in a meaningful context.

2 More Depth of Coverage

Science textbooks can be overwhelming for many children, especially those who have reading problems. They often contain unfamiliar vocabulary and tend to cover a broad range of topics (Casteel and Isom 1994; Short and Armstrong 1993; Tyson and Woodward 1989). However, fiction and non-fiction picture books tend to focus on fewer topics

and give more in-depth coverage of the concepts. It can be useful to pair an engaging fiction book with a nonfiction book to round out the science content being presented.

For example, "Be a Friend to Trees," the Chapter 12 lesson, features *Our Tree Named Steve,* a poignant story of a father's recounting memories of the family's special tree. It is paired with *Be a Friend to Trees,* a nonfiction book that explains the importance of trees as sources of food, oxygen, and other essential things. The emotion-engaging storyline in *Our Tree Named Steve* hooks the reader, and the book *Be a Friend to Trees* presents facts and background information. Together they offer a balanced, in-depth look at how trees are important to people and animals.

3 Improved Reading and Science Skills

Research by Morrow et al. (1997) on using children's literature and literacy instruction in the science program indicated gains in science as well as literacy. Romance and Vitale (1992) found significant improvement in both science and reading scores of fourth graders when the regular basal reading program was replaced with reading in science that correlated with the science curriculum. They also found an improvement in students' attitudes toward the study of science.

4 Opportunities to Correct Science Misconceptions

Students often have strongly held misconceptions about science that can interfere with their learning. "Misconceptions, in the field of science education, are preconceived ideas that differ from those currently accepted by the scientific community" (Colburn 2003, p. 59). Children's picture books, reinforced with hands-on inquiries, can help students correct their misconceptions. Repetition of the correct concept by reading several books, doing a number of experiments, and inviting scientists to the classroom can facilitate a conceptual change in children (Miller, Steiner, and Larson 1996).

But teachers must be aware that scientific misconceptions can be inherent in the picture books. Although many errors are explicit, some of the misinformation is more implicit or may be inferred from text and illustrations (Rice 2002). This problem is more likely to occur within fictionalized material. Mayer's (1995) study demonstrated that when both inaccuracies and science facts are presented in the same book, children do not necessarily remember the correct information.

Scientific inaccuracies in picture books can be useful for teaching. Research shows that errors in picture books, whether identified by the teacher or the students, can be used to help children learn to question the accuracy of what they read by comparing their own observations to the science presented in the books (Martin 1997). Scientifically inaccurate children's books can be helpful when students analyze inaccurate text or pictures after they have gained understanding of the correct scientific concepts through inquiry experiences.

For example, in the "That Magnetic Dog" lesson, Chapter 13, after using magnets and reading a nonfiction book about magnets, students analyze an inaccurate sentence in the book *That Magnetic Dog* and then rewrite the sentence in a way that is scientifically correct. This process requires students to think critically: They apply what they have learned to evaluate and correct the misinformation in the picture book.

Selection of Books

Each lesson in *More Picture-Perfect Science Lessons* focuses on one or more of the National Science Education Standards. We selected fiction and nonfiction children's picture books that closely relate to the Standards. An annotated "More Books to Read" section is provided at the end of each lesson. If you would like to select more children's literature to use in your science classroom, try *The Outstanding Science Trade Books for Students K–12* listing, a cooperative project between the National Science Teachers Association (NSTA) and the Children's Book Council (CBC). The books are selected by a book review panel appointed by the NSTA and assembled in cooperation with the CBC. Each year a new list is featured in the March issue of NSTA's elementary school teacher journal *Science and Children*. See *www.nsta.org/ostbc* for archived lists.

When you select children's picture books for science instruction, you should consult with a knowledgeable colleague who can help you check them for errors or misinformation. You might talk with a high school science teacher, a retired science teacher, or a university professor. To make sure that the books are developmentally appropriate or lend themselves to a particular reading strategy you want to model, you could consult with a language arts specialist.

Finding Out-of-Print Books

We have included the most up-to-date information we have, but children's picture books go in and out of print frequently. Check your school library, public library, or a used-book store for copies of out-of-print books. In addition, the following websites may be helpful:

- *www.abebooks.com*—abebooks.com is a large online marketplace for books that can locate new, used, rare, or out-of-print books through a community of more than 12,000 independent booksellers from around the world.

- *www.alibris.com*—Alibris connects people with books, music, and movies from thousands of independent sellers around the world. They offer more than 35 million used, new, and hard-to-find titles to consumers, libraries, and retailers.

- *www.bibliofind.com*—Bibliofind has combined with Amazon.com to provide millions of rare, used, and out-of-print books.

- *www.powells.com*—Powell's has an extensive list of both new and used books.

Considering Genre

Considering genre when you determine how to use a particular picture book within a science lesson is important. Donovan and Smolkin (2002) identify four different genres frequently recommended for teachers to use in their science instruction: story, non-narrative information, narrative information, and dual purpose. *More Picture-Perfect Science Lessons* identifies the genre of each featured book at the beginning of each lesson. Summaries of the four genres, a representative picture book for each genre, and suggestions for using each genre within the BSCS 5E learning cycle we use follow. (The science learning cycle known as the BSCS 5E Model is described in detail in Chapter 4.)

Storybooks

Storybooks center on specific characters who work to resolve a conflict or problem. The major purpose of stories is to entertain, not to present factual information. The vocabulary is typically commonsense, everyday language. An engaging storybook can spark interest in a science topic and move students toward informational texts to answer questions inspired by the story. For example, "Bubbles," Chapter 6, uses Mercer Mayer's *Bubble, Bubble,* a story about a boy who buys a magic bubble maker that blows bubbles in the shapes of animals. The imaginative story hooks the learners and engages them in an investigation to find out if free-floating bubbles can really be different shapes or if they are always round.

Scientific concepts in stories are often implicit, so teachers must make the concepts explicit to students. As we mentioned, be aware that storybooks often contain scientific errors, either explicit or implied by text or illustrations. Storybooks with scientific errors can be used toward the end of a lesson to teach students how to identify and correct the inaccurate science. For example, "That Magnetic Dog," Chapter 13, features Bruce Whatley's *That Magnetic Dog,* a storybook that contains some scientific inaccuracies. Books like this can be powerful vehicles for assessing the ability of learners to analyze the scientific accuracy of a text.

Non-narrative Information Books

Non-narrative information books are factual texts that introduce a topic, describe the attributes of the topic, or describe typical events that occur. The focus of these texts is on the subject matter, not specific characters. The vocabulary is typically technical. Readers can enter the text at any point in the book. Many contain features found in nonfiction such as a table of contents, bold-print vocabulary words, a glossary, and an index. Young children tend to be less familiar with this genre and need many opportunities to experience this type of text. Using non-narrative information books will help students become familiar with the structure of textbooks, as well as "real-world" reading, which is primarily nonfiction. Teachers may want to read only those sections that provide the concepts and facts needed to meet particular science objectives.

We wrote the articles included in some of the lessons (see chapters 4 and 12) in non-narrative information style to give students more opportunity to practice reading this type of text. For example, "Loco Beans," Chapter 9, includes an article written in an expository style that shows key words in bold print. Another example of non-narrative information writing is the book *Coral Reef Animals,* which contains nonfiction text features such as a table of contents, bold-print words, insets, a glossary, and an index. *Coral Reef Animals* is featured in "Over in the Ocean," Chapter 11. The appropriate placement of non-narrative information text in a science learning cycle is after students have had the opportunity to explore concepts through hands-on activities. At that point, students are engaged in the topic and are motivated to read the non-narrative informational text to learn more.

Narrative Information Books

Narrative information books, sometimes referred to as "hybrid books," provide an engaging format for factual information. They communicate a sequence of factual events over time and sometimes

recount the events of a specific case to generalize to all cases. When using these books within science instruction, establish a purpose for reading so that students focus on the science content rather than the storyline. In some cases, teachers may want to read the book one time through for the aesthetic components of the book and a second time for specific science content. *Rachel Carson: Preserving a Sense of Wonder*, an example of a narrative information text, is used in "A Sense of Wonder," Chapter 20. This narrative chronicles the life and legacy of groundbreaking environmentalist Rachel Carson. The narrative information genre can be used at any point within a science learning cycle. This genre can be both engaging and informative.

Dual-Purpose Books

Dual-purpose books are intended to serve two purposes: present a story and provide facts. They employ a format that allows readers to use the book like a storybook or to use it like a non-narrative information book. Sometimes information can be found in the running text, but more frequently it appears in insets and diagrams. Readers can enter on any page to access specific facts, or they can read the book through as a story. You can use the story component of a dual-purpose book to engage the reader at the beginning of the science learning cycle. For example, Chapter 9 features the book, *Lucas and His Loco Beans*, which is used to engage the students in an investigation of the life cycle of the Mexican Jumping Bean Moth.

Dual-purpose books typically have little science content within the story. Most of the informational ideas are found in the insets and diagrams. If the insets and diagrams are read, discussed, explained, and related to the story, these books can be very useful in helping students refine concepts and acquire scientific vocabulary *after* they have had opportunities for hands-on exploration. *Imaginative Inventions* is a dual-purpose book featured in Chapter 19. Each page contains a humorous poem about an invention with insets on the edge of the page that list facts about the invention.

Using Fiction and Nonfiction Texts

As we mentioned previously, pairing fiction and nonfiction books in read alouds to round out the science content being presented can be useful. Because fiction books tend to be very engaging for students, they can be used to hook students at the beginning of a science lesson. But most of the reading people do in everyday life is nonfiction. We are immersed in informational text every day, and we must be able to comprehend it in order to be successful in school, at work, and in society. Nonfiction books and other informational text such as articles should be used frequently in the elementary classroom. They often include text structures that differ from stories, and the opportunity to experience these structures in read alouds can strengthen students' abilities to read and understand informational text. Duke (2004) recommends four strategies to help teachers improve students' comprehension of informational text. Teachers should

- increase students' access to informational text,

- increase the time they spend working with informational text,

- teach comprehension strategies through direct instruction, and

- create opportunities for students to use informational text for authentic purposes.

More Picture-Perfect Science Lessons addresses these recommendations in several ways. The lessons expose students to a variety of nonfiction picture books and articles on science topics, thereby increasing access to informational text. The lessons explain how anticipation guides, pairs reading, and using nonfiction features all help improve students' comprehension of the informational text by increasing the time they spend working with it. Each lesson also includes instructions for explicitly teaching comprehension strategies within the learning cycle. The inquiry-based lessons provide an authentic purpose for reading informational

text, as students are motivated to read or listen in order to find the answers to questions generated within the inquiry activities.

References

Butzow, J., and C. Butzow. 2000. *Science through children's literature: An integrated approach*. Portsmouth, NH: Teacher Ideas Press.

Casteel, C. P., and B. A. Isom. 1994. Reciprocal processes in science and literacy learning. *The Reading Teacher* 47: 538–544.

Colburn, A. 2003. *The lingo of learning: 88 education terms every science teacher should know*. Arlington, VA: NSTA Press.

Donovan, C., and L. Smolkin. 2002. Considering genre, content, and visual features in the selection of trade books for science instruction. *The Reading Teacher* 55: 502–520.

Duke, N. K. 2004. The case for informational text. *Educational Leadership* 61: 40–44.

Harvey, S., and A. Goudvis. 2000. *Strategies that work: Teaching comprehension to enhance understanding*. York, ME: Stenhouse Publishers.

Kralina, L. 1993. Tricks of the trades: Supplementing your science texts. *The Science Teacher* 60(9): 33–37.

Martin, D. J. 1997. *Elementary science methods: A constructivist approach*. Albany, NY: Delmar.

Mayer, D. A. 1995. How can we best use children's literature in teaching science concepts? *Science and Children* 32(6): 16–19, 43.

Miller, K. W., S. F. Steiner, and C. D. Larson. 1996. Strategies for science learning. *Science and Children* 33(6): 24–27.

Morrow, L. M., M. Pressley, J. K. Smith, and M. Smith. 1997. The effect of a literature-based program integrated into literacy and science instruction with children from diverse backgrounds. *Reading Research Quarterly* 32: 54–76.

National Research Council. 1996. *National Science Education Standards*. Washington, DC: National Academy Press. Available online at *books.nap. edu/books/0309053269/html/index.html*

Rice, D. C. 2002. Using trade books in teaching elementary science: Facts and fallacies. *The Reading Teacher* 55(6): 552–565.

Romance, N. R., and M. R. Vitale. 1992. A curriculum strategy that expands time for in-depth elementary science instruction by using science-based reading strategies: Effects of a year-long study in grade four. *Journal of Research in Science Teaching* 29: 545–554.

Short, K. G., and J. Armstrong. 1993. Moving toward inquiry: Integrating literature into the science curriculum. *New Advocate* 6(3): 183–200.

Strickland, D. S., and L. M. Morrow, eds. 2000. *Beginning reading and writing*. New York: Teachers College Press.

Tyson, H., and A. Woodward. 1989. Why aren't students learning very much from textbooks? *Educational Leadership* 47(3): 14–17.

Children's Books Cited

Aliki. 1981. *Digging up dinosaurs*. New York: Harper-Trophy.

Carle, E. 1981. *The very hungry caterpillar*. New York: Philomel.

Cleary, B. 1968. *Ramona the pest*. New York: HarperCollins.

Galko, F. 2003. *Coral reef animals*. Chicago: Heinemann Library.

Harper, C. M. 2001. *Imaginative inventions*. New York: Little, Brown and Company.

Lauber, P. 1994. *Be a friend to trees*. New York: Harper-Trophy.

Lobel, A. 1979. *Frog and toad are friends*. New York: HarperTrophy.

Locker, T., and J. Bruchac. 2004. *Rachel Carson: Preserving a sense of wonder*. Golden, CO: Fulcrum Publishing .

Mayer, M. 1973. *Bubble bubble*. Columbus, OH: Gingham Dog Press.

Rey, H. A. 1973. *Curious George*. Boston: Houghton Mifflin.

Sendak, M. 1988. *Where the wild things are*. New York: HarperCollins.

Seuss, Dr. 1960. *One fish two fish red fish blue fish*. New York: Random House Books for Young Readers

Silverstein, S. 1974. *Where the sidewalk ends*. New York: HarperCollins.

Simon, S. 1984. *The Moon*. Salem, OR: Four Winds.

Stone, J. 2003. *The monster at the end of this book*. New York: Golden Books.

Whatley, B. 1994. *That magnetic dog*. Sydney, Australia: Angus & Robertson.

White, E. B. 1952. *Charlotte's web*. New York: Harper-Collins.

Winner, R. M. 2002. *Lucas and his loco beans*. Santa Barbara, CA: Brainstorm 3000.

Zweibel, A. 2005. *Our tree named Steve*. New York: G.P. Putnam's Sons.

Reading Aloud

This chapter addresses some of the research supporting the importance of reading aloud, tips to make your read-aloud time more valuable, descriptions of Harvey and Goudvis's six key reading strategies (2000), and tools you can use to enhance students' comprehension during read-aloud time.

Why Read Aloud?

Being read to is the most influential element in building the knowledge required for eventual success in reading (Anderson et al. 1985). It improves reading skills, increases interest in reading and literature, and can even improve overall academic achievement. A good reader demonstrates fluent, expressive reading and models the thinking strategies of proficient readers, helping to build background knowledge and fine-tune students' listening skills. When a teacher does the reading, children's minds are free to anticipate, infer, connect, question, and comprehend (Calkins 2000). In addition, being read to is risk-free. In *Yellow Brick Roads: Shared and Guided Paths to Independent Reading 4–12* (2000, p. 45), Allen says, "For

Read-aloud time is a special part of Mrs. Slusser's class

students who struggle with word-by-word reading, experiencing the whole story can finally give them a sense of the wonder and magic of a book."

Reading aloud is appropriate in all grade levels and for all subjects. It is important not only when children can't read on their own but also after they have become proficient readers (Anderson et al. 1985). Allen supports this view: "Given the body of research supporting the importance of read-aloud for modeling fluency, building background knowledge, and developing language acquisition, we should remind ourselves that those same benefits occur when we extend read-aloud beyond the early years" (2000, p. 44).

Ten Tips for Reading Aloud

We have provided a list of tips to help you get the most from your read-aloud time. Using these suggestions can help set the stage for learning, improve comprehension of science material, and make the read-aloud experience richer and more meaningful for both you and your students.

1 Preview the Book

Select a book that meets your science objectives *and* lends itself to reading aloud. Preview it carefully before sharing it with the students. Are there any errors in scientific concepts or misinformation that could be inferred from the text or illustrations? If the book is not in story form, is there any nonessential information you could omit to make the read-aloud experience better? If you are not going to read the whole book, choose appropriate starting and stopping points before reading. Consider generating questions and inferences about the book in advance and placing them on sticky notes inside the book to help you model your thought processes as you read aloud.

2 Set the Stage

Because reading aloud is a performance, you should pay attention to the atmosphere and physical setting of the session. Gather the students in a special reading area, such as on a carpet or in a semicircle of chairs. Seat yourself slightly above them. Do not sit in front of a bright window where the glare will keep students from seeing you well or in an area where students can be easily distracted. You may want to turn off the overhead lights and read by the light of a lamp or use soft music as a way to draw students into the mood of the text. Establish expectations for appropriate behavior during read-aloud time, and, before reading, give the students an opportunity to settle down and focus their attention on the book.

3 Celebrate the Author and Illustrator

Always announce the title of the book, the author, and the illustrator before reading. Build connections by asking students if they have read other books by the author or illustrator. Increase interest by sharing facts about the author or illustrator from the book's dust jacket or from library or internet research. This could be done either before or after the reading. The following resources are useful for finding information on authors and illustrators:

Books

- Kovacs, D., and J. Preller. 1991. *Meet the authors and illustrators: Volume one.* New York: Scholastic.

- Kovacs, D., and J. Preller. 1993. *Meet the authors and illustrators: Volume two.* New York: Scholastic.

- Peacock, S. 2003. *Something about the author: Facts and pictures about authors and illustrators of books for young people.* Farmington Hills, MI: Gale Group.

- Preller, J. 2001. *The big book of picture-book authors and illustrators.* New York: Scholastic.

Websites

- *www.teachingbooks.net*—Teaching Books continually identifies, catalogs, and maintains reliable links to children's books' author and illustrator websites and organizes them into categories relevant to teachers' needs.

- *www.cbcbooks.org*—The Children's Book Council (CBC) is a nonprofit trade organization encouraging literacy and the use and enjoyment of children's books. Its website has a feature titled "About Authors and Illustrators" with links to author and illustrator websites.

4 Read With Expression

Practice reading aloud to improve your performance. Can you read with more expression to more fully engage your audience? Try louder or softer speech, funny voices, facial expressions, or gestures. Make eye contact with your students every now and then as you read. This strengthens the bond between reader and listener, helps you gauge your audience's response, and cuts down on off-task behaviors. Read slowly enough that your students have time to build mental images of what you are reading, but not so slowly that they lose interest. When reading a nonfiction book aloud, you may want to pause after reading about a key concept to let it sink in and then reread that part. At suspenseful parts in a storybook, use dramatic pauses or slow down and read softly. This can move the audience to the edge of their seats.

5 Share the Pictures

Don't forget the power of visual images to help students connect with and comprehend what you are reading. Make sure that you hold the book in a way such that students can see the pictures on each page. Read captions if appropriate. In some cases, you may want to hide certain pictures so students can visualize what is happening in the text before you reveal the illustrator's interpretation.

6 Encourage Interaction

Keep chart paper and markers nearby in case you want to record questions or new information. Try providing students with "think pads" in the form of sticky notes to write on as you read aloud. Not only does this help extremely active children keep their hands busy while listening, but it also encourages students to interact with the text as they jot down questions or comments. After the read aloud, have students share

their questions and comments. You may want students to place their sticky notes on a class chart whose subject is the topic being studied. Another way to encourage interaction without taking the time for each student to ask questions or comment is to do an occasional "turn and talk" during the read aloud. Stop reading, ask a question, allow thinking time, and then have each student share ideas with a partner.

7 Keep the Flow

Although you want to encourage interaction during a read aloud, avoid excessive interruptions that may disrupt fluent, expressive reading. Aim for a balance between allowing students to hear the language of the book uninterrupted and providing them with opportunities to make comments, ask questions, and share connections to the reading. As we have suggested, you may want to read the book all the way through one time so students can enjoy the aesthetic components of the story. Then go back and read the book for the purpose of meeting the science objectives.

8 Model Reading Strategies

As you read aloud, it is important that you help children access what they already know and build bridges to new understandings. Think out loud, model your questions for the author, and make connections to yourself, other books, and the world. Show students how to determine the important parts of the text or story, and demonstrate how you synthesize meaning from the text. Modeling these reading comprehension strategies when appropriate before, during, and/or after reading helps students internalize the strategies and begin to use them in their own reading. Six key strategies are described in detail later in this chapter.

9 Don't Put It Away

Keep the read-aloud book accessible to students after you read it. They will want to get a close-up look at the pictures and will enjoy reading the book independently. Don't be afraid of reading the same book more than once—children benefit from the repetition.

10 Have Fun

Let your passion for books show. It is contagious! Read nonfiction books with interest and wonder. Share your thoughts, question the author's intent, synthesize meaning out loud, and voice your own connections to the text. When reading a story, let your emotions show—laugh at the funny parts and cry at the sad parts. Seeing an authentic response from the reader is important for students. If you read with enthusiasm, read-aloud time will become special and enjoyable for everyone involved.

We hope these tips will help you and your students reap the many benefits of read alouds. As Debbie Miller writes in *Reading with Meaning: Teaching Comprehension in the Primary Grades* (2002, p. 26), "Learning to read should be a joyful experience. Give children the luxury of listening to well-written stories with interesting plots, singing songs and playing with their words, and exploring a wide range of fiction, nonfiction, poetry and rhymes…. Be genuine. Laugh. Love. Be patient. You're creating a community of readers and thinkers."

Reading Comprehension Strategies

Children's author Madeleine L'Engle says, "Readers usually grossly underestimate their own importance. If a reader cannot create a book along with the writer, the book will never come to life. The author and the reader … meet on the bridge of words." (1995, p. 34). It is our responsibility as teachers, no matter what subjects we are assigned to teach, to help children realize the importance of their own thoughts and ideas as they read. Modeling our own thinking as we read aloud is the first step. Becoming a proficient reader is an ongoing, complex process, and children need to be explicitly taught the strategies that good readers use. In *Strategies that Work*, Harvey and Goudvis identify six key reading strategies essential to achieving full understanding when we read. These strategies are used where appropriate in each lesson and are seamlessly embedded into the 5E model. The strategies should be modeled as you read aloud to students from both fiction and nonfiction texts.

Research shows that explicit teaching of reading comprehension strategies can foster comprehension development (Duke and Pearson 2002). Explicit teaching of the strategies is the initial step in the gradual-release-of-responsibility approach to delivering reading instruction (Fielding and Pearson 1994). During this first phase of the gradual-release method, the teacher *explains* the strategy, demonstrates *how* and *when* to use the strategy, explains *why* it is worth using, and *thinks aloud* in order to model the mental processes used by good readers. Duke (2004, p. 42) describes this process: "I often discuss the strategies in terms of good readers, as in 'Good readers think about what might be coming next.' I also model the uses of comprehension strategies by thinking aloud as I read. For example, to model the importance of monitoring understanding, I make comments such as, 'That doesn't make sense to me because …' or 'I didn't understand that last part—I'd better go back.'" Using the teacher-modeling phase within a science learning cycle will reinforce what students do during reading instruction, when the gradual-release-of-responsibility model can be continued. When students have truly mastered a strategy, they are able to apply it to a variety of texts and curricular areas and can explain how the strategy helps them construct meaning.

Descriptions of the six key reading comprehension strategies featured in *Strategies that Work* follow. The ⛏ icon highlights these strategies here and within the lessons.

⛏ Making Connections

Making meaningful connections during reading can improve learners' comprehension and engagement by helping them better relate to what they read. Comprehension breakdown that occurs when reading or listening to expository text can come from a lack of prior information. These three techniques can help readers build background knowledge where little exists.

Asking questions is not only a critical reading skill but is also at the heart of scientific inquiry and can lead students into meaningful investigations.

Visualizing

Visualizing is the creation of mental images while reading or listening to text. Mental images are created from the learner's emotions and senses, making the text more concrete and memorable. Imagining the sensory qualities of things described in a text can help engage learners and stimulate their interest in the reading. When readers form pictures in their minds, they are also more likely to stick with a challenging text. During a reading, you can stop and ask students to visualize the scene. What sights, sounds, smells, colors are they imagining?

Applying the questioning strategy

- *Text-to-Self* connections occur when readers and listeners link the text to their past experiences or background knowledge.

- *Text-to-Text* connections occur when readers and listeners recognize connections from one book to another.

- *Text-to-World* connections occur when readers and listeners connect the text to events or issues in the real world.

Questioning

Proficient readers ask themselves questions before, during, and after reading. Questioning allows readers to construct meaning, find answers, solve problems, and eliminate confusion as they read. It motivates readers to move forward in the text. In *Strategies that Work* (2000, p. 82), Harvey and Goudvis write, "A reader with no questions might just as well abandon the book. When our students ask questions and search for answers, we know that they are monitoring comprehension and interacting with the text to construct meaning, which is exactly what we hope for in developing readers."

Visualizing a roller coaster ride

Inferring

Reading between the lines, or inferring, involves a learner's merging clues from the reading with prior knowledge to draw conclusions and interpret the text. Good readers make inferences before, during, and after reading. Inferential thinking is also

Making meaning

Tools to Enhance Comprehension

We have identified several activities and organizers that can enhance students' science understanding and reading comprehension in the lessons. These tools, which support the Harvey and Goudvis reading comprehension strategies, are briefly described on the following pages and in more detail within the lessons.

Anticipation Guides

Anticipation guides (Herber 1978) are sets of questions that serve as a pre- or postreading activity for a text. They can be used to activate and assess prior knowledge, determine misconceptions, focus thinking on the reading, and motivate reluctant readers by stimulating interest in the topic. An anticipation guide should revolve around four to six key concepts from the reading that learners respond to before reading. They will be motivated to read or listen carefully to find the evidence that supports their predictions. After reading, learners revisit their anticipation guide to check their responses. In a revised extended anticipation guide (Duffelmeyer and Baum 1992), learners are required to justify their responses and explain why their choices were correct or incorrect.

Chunking

Chunking is dividing the text into manageable sections and reading only a section at any one time. This gives learners time to digest the information in a section before moving on. Chunking is also a useful technique for weeding out essential from nonessential information when reading nonfiction books. Reading only those parts of the text that meet your learning objectives focuses the learning on what is important. Remember—nowhere is it written that you must read nonfiction books cover to cover when doing a read aloud. Feel free to omit parts that are inaccurate, out of date, or don't contribute in a meaningful way to the lesson.

an important science skill and can be reinforced during reading instruction.

Determining Importance

Reading to learn requires readers to identify essential information by distinguishing it from nonessential details. Deciding what is important in the text depends upon the purpose for reading. In *More Picture-Perfect Science Lessons,* the lesson's science objectives determine importance. Learners read or listen to the text to find answers to specific questions, to gain understanding of science concepts, and to identify science misconceptions.

Synthesizing

In synthesizing, readers combine information gained through reading with prior knowledge and experience to form new ideas. To synthesize, readers must stop, think about what they have read, and contemplate its meaning before continuing on through the text. The highest level of synthesis involves those "aha!" moments when readers achieve new insight and, as a result, change their thinking.

Cloze Strategy

Cloze refers to an activity that helps readers infer the meanings of unfamiliar words. In the cloze strategy, key words are deleted in a passage. Students then fill in the blanks with words that make sense and sound right.

Most Valuable Point

Most valuable point, or MVP, is a tool for determining importance. After reading a text or passage, students are asked to determine the MVP. Students can be encouraged to write VIPs (*very important points*), but must decide on one MVP. The purpose of this activity is for students to learn how to distinguish between important and unimportant information so they can identify key ideas as they read.

O-W-L Chart

An *O-W-L*—**O**bservations, **W**onderings and **L**earnings—(Ansberry and Morgan 2005) *chart* is a three-column chart in which students record their observations about a phenomenon or object, their wonderings about it, and what they learn about it. An O-W-L chart can help a reader focus on certain information based on their wonderings and provides a place for students to record what they learned from the reading. After reading, students can add more questions to the "wonderings" column of the chart.

Pairs Read

Pairs read (Billmeyer and Barton 1998) requires learners to work cooperatively as they read and make sense of a text. While one learner reads aloud, the other listens and then summarizes the main idea. Encourage students to ask their partners to re-read if they need clarification. Benefits of pairs read include increased reader involvement, attention, and collaboration. In addition, students become more independent, less reliant on the teacher.

Picture Walk

A *picture walk* consists of showing students the cover of a book and browsing through the pages,

Using an O-W-L chart

in order, without reading the text. The purpose of this tool is to establish interest in the story and expectations about what is to come. It also reinforces the importance of using visual cues while reading. Students look at the pictures and talk about what they see, what may be happening in each illustration, and how the pictures come together to make a story. Some useful questions to ask during a picture walk are

- From looking at the cover, what do you think this book is about?

- What do you see?

- What do you think is happening?

- What do you think will happen next?

- What are you curious to know more about in the story?

Pairs read

Questioning the Author

Questioning the author, or QtA (Beck et al. 1997) is an interactive strategy that helps students comprehend what they are reading. When students read in a QtA lesson, they learn to question the ideas presented in the text while they are reading, making them critical thinkers, not just readers. This strategy can be very effective in the science classroom as a way to keep students from acquiring misconceptions from the text and/or illustrations in a picture book or textbook.

Rereading

Nonfiction text is often full of unfamiliar ideas and difficult vocabulary. *Rereading* content for clarification is an essential skill of proficient readers, and you should model this frequently. Rereading content for a different purpose can aid comprehension. For example, you might read aloud a text for enjoyment and then revisit the text to focus on specific science content.

Sketch to Stretch

During *sketch to stretch* (Seigel 1984), learners pause to reflect on the text and do a comprehension self-assessment by drawing on paper the images they visualize in their heads during reading. They might illustrate an important event from the text, sketch the characters in a story, or make a labeled diagram. Have students use pencils so they understand the focus should be on collecting their thoughts rather than creating a piece

of art. You may want to use a timer so students understand that sketch to stretch is a brief pause to reflect quickly on the reading. Students can share and explain their drawings in small groups after sketching.

Stop and Jot

In *stop and jot*, learners stop and think about the reading and then jot down a thought. They may write about something they've just learned, something they are wondering about, or what they expect to learn next. If they use sticky notes for this, the notes can be added to a whole-class chart to connect past and future learning.

Turn and Talk

Learners pair up with a partner to share their ideas, explain concepts in their own words, or tell about a connection they have to the book in *turn and talk*. Sometimes called *think-pair-share* or *sharing pairs*, this strategy allows each child to respond so that everyone in the group is involved as either a talker or a listener. Saying "Take a moment to turn and talk about your ideas with someone" gives students an opportunity to satisfy their needs to express their own thoughts about the reading.

Using Features of Nonfiction

Many nonfiction books include a table of contents, index, glossary, bold-print words, picture cap-

Taking a picture walk

Stop and jot

tions, diagrams, and charts that provide valuable information. Because children are generally more used to narrative text, they often skip over these text structures. It is important to model how to interpret the information these features provide the reader. To begin, show the cover of a nonfiction book and read the title and table of contents. Ask students to predict what they'll find in the book. Show students how to use the index in the back of the book to find specific information. Point out other nonfiction text structures as you read and note that these features are unique to nonfiction. Model how nonfiction books can be entered at any point in the text, because they generally don't follow a storyline.

Venn Diagram

A *Venn diagram* is made of two or more overlapping circles and is useful for comparing two or more items, books, people, animals, events, or anything else you wish to compare. Each circle is labeled. Similarities are written in the space where the circles intersect. Differences are written in the parts of the circle that do not intersect. This tool

can help students organize their thinking and sort through information after reading.

Why Do Picture Books Enhance Comprehension?

Students should be encouraged to read a wide range of print materials, but picture books offer many advantages when teaching reading-comprehension strategies. Harvey and Goudvis not only believe that interest is essential to comprehension, but they also maintain that, because picture books are extremely effective for building background knowledge and teaching content, instruction in reading comprehension strategies during picture book read alouds allows students to better access that content. In summary, picture books are invaluable for teaching reading-comprehension strategies because they are extraordinarily effective at keeping readers engaged and thinking.

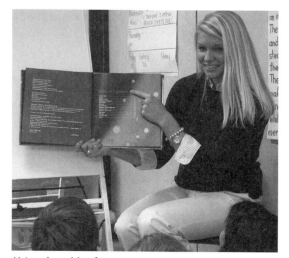
Using the table of contents

References

Allen, J. 2000. *Yellow brick roads: Shared and guided paths to independent reading 4–12.* Portland, ME: Stenhouse Publishers.

Anderson, R. C., E. H. Heibert, J. Scott, and I. A. G. Wilkinson. 1985. *Becoming a nation of readers: The report of the commission on reading.* Champaign, IL: Center for the Study of Reading; Washington, DC: National Institute of Education.

Using a Venn diagram

Duke, N. K. 2004. The case for informational text. *Educational Leadership* 61: 40–44.

Duke, N. K., and P. D. Pearson. 2002. Effective practices for developing reading comprehension. In *What research has to say about reading instruction,* edited by A. E. Farstrup and S. J. Samuels. Newark, DE: International Reading Association.

Fielding, L., and P. D. Pearson. 1994. Reading comprehension: What works? *Educational Leadership* 51(5): 62–67.

Gillett, J. W., and C. Temple. 1983. *Understanding reading problems: Assessment and instruction.* Boston: Little, Brown.

Harvey, S., and A. Goudvis. 2000. *Strategies that work: Teaching comprehension to enhance understanding.* York, ME: Stenhouse Publishers.

Herber, H. 1978. *Teaching reading in the content areas.* Englewood Cliffs, NJ: Prentice Hall.

L'Engle, M. 1995. *Walking on water: Reflections on faith and art.* New York: North Point Press.

Miller, D. 2002. *Reading with meaning: Teaching comprehension in the primary grades.* Portland, ME: Stenhouse Publishers.

Seigel, M. 1984. Sketch to stretch. In *Reading, writing, and caring,* edited by O. Cochran. New York: Richard C. Owen.

Ansberry, K. R., and E. Morgan. 2005. *Picture-perfect science lessons: Using children's books to guide inquiry.* Arlington, VA: NSTA Press.

Beck, I., R. Hamilton, L. Kucan, and M. McKeown. 1997. *Questioning the author: An approach for enhancing student engagement with text.* Newark, DE: International Reading Association.

Beers, S., and L. Howell. 2004. *Reading strategies for the content areas: An action tool kit, volume 2.* Alexandria, VA: Association for Supervision and Curriculum Development.

Billmeyer, R., and M. L. Barton. 1998. *Teaching reading in the content areas: If not me, then who?* Aurora, CO: Mid-continent Regional Educational Leadership Laboratory.

Calkins, L. M. 2000. *The art of teaching reading.* Boston: Pearson Allyn & Bacon.

Duffelmeyer, F. A., and D. D. Baum. 1992. The extended anticipation guide revisited. *Journal of Reading* 35: 654–656.

Teaching Science Through Inquiry

The word *inquiry* brings many different ideas to mind. For some teachers, it may evoke fears of giving up control in the classroom or spending countless hours preparing lessons. For others, it may imply losing the focus of instructional objectives while students pursue answers to their own questions. And for many, teaching science through inquiry is perceived as intriguing but unrealistic. But inquiry doesn't have to cause anxiety for teachers. Simply stated, inquiry is an approach to learning that involves exploring the world and that leads to asking questions, testing ideas, and making discoveries in the search for understanding. There are many degrees of inquiry, and it may be helpful to start with a variation that emphasizes a teacher-directed approach and then gradually builds to a more student-directed approach. As a basic guide, the National Research Council (2000) identifies five essential features for classroom inquiry, shown in Table 3.1.

Essential Features of Classroom Inquiry

The following descriptions illustrate each of the five essential features of classroom inquiry (NRC 2000) using Chapter 13, "That Magnetic Dog." Any classroom activity that includes all five of these features is considered to be inquiry.

1 *Learners are engaged by scientifically oriented questions.* In *That Magnetic Dog,* students are engaged in the question "What is special about objects that are attracted to a magnet?"

2 *Learners give priority to evidence, which allows them to develop and evaluate explanations that address scientifically oriented questions.* Students

---

Table 3.1

Five Essential Features of Classroom Inquiry

1 Learners are engaged by scientifically oriented questions.

2 Learners give priority to evidence, which allows them to develop and evaluate explanations that address scientifically oriented questions.

3 Learners formulate explanations from evidence to address scientifically oriented questions.

4 Learners evaluate their explanations in light of alternative explanations, particularly those reflecting scientific understanding.

5 Learners communicate and justify their proposed explanations.

From *Inquiry and the National Science Education Standards: A Guide for Teaching and Learning* (NRC 2000).

---

collect evidence by using a magnet to "fish" for items in a box. They put all of the items attracted to the magnet in a pile outside of the box, and they leave the items that were not attracted to the magnet in the box.

3 *Learners formulate explanations from evidence to address scientifically oriented questions.* Students compare the items that were attracted to a magnet to the items that were not attracted to the magnet and develop hypotheses about why the items were attracted to the magnet.

4 *Learners evaluate their explanations in light of alternative explanations, particularly those reflecting scientific understanding.* Students consult a nonfiction book about magnets to find out if their hypotheses were correct.

5 *Learners communicate and justify their proposed explanations.* Students create posters displaying magnetic and nonmagnetic items and use the posters to help explain their conclusions.

Benefits of Inquiry

Developing an inquiry-based science program is a central tenet of the *National Science Education Standards* (NRC 1996). So what makes inquiry-based teaching such a valuable method of instruction? Many studies state that it is equal or superior to other instructional modes and results in higher scores on content achievement tests. *Inquiry and the National Science Education Standards* (NRC 2000) summarizes the findings of *How People Learn* (Bransford, Brown, and Cocking 1999), which support the use of inquiry-based teaching. Those findings include the following points:

- Understanding science is more than knowing facts. Most important is that students understand the major concepts. Inquiry-based teaching focuses on the major concepts, helps students build a strong base of factual information to support the concepts, and gives them opportunities to apply their knowledge effectively.

- Students build new knowledge and understanding on what they already know and believe. Students often hold preconceptions that either are reasonable in only a limited context or are scientifically incorrect. These preconceptions can be resistant to change, particularly when teachers use conventional teaching strategies (Wandersee, Mintzes, and Novak 1994). Inquiry-based teaching uncovers students' prior knowledge and, through concrete explorations, helps them correct misconceptions.

- Students formulate new knowledge by modifying and refining their current concepts and by adding new concepts to what they already know. In an inquiry-based model, students give priority to evidence when they prove or disprove their preconceptions. Their preconceptions are challenged by their observations or the explanations of other students.

- Learning is mediated by the social environment in which learners interact with others. Inquiry provides students with opportunities to interact with others. They explain their ideas to other students and listen critically to the ideas of their classmates. These social interactions require that students clarify their ideas and consider alternative explanations.

- Effective learning requires that students take control of their own learning. When teachers use inquiry, students assume much of the responsibility for their own learning. Students formulate questions, design procedures, develop explanations, and devise ways to share their findings. This makes learning unique and more valuable to each student.

- The ability to apply knowledge to novel situations, that is, transfer of learning, is affected by the degree to which students learn with understanding. Inquiry provides students a variety of opportunities to practice what they have learned, connect to what they already know, and therefore moves them toward application, a sophisticated level of thinking that requires them to solve problems in new situations.

Inquiry learning not only contributes to better understanding of scientific concepts and skills, but, because science inquiry in school is carried out in a social context, it also contributes to children's social and intellectual development (Dyasi 1999). Within an inquiry-based lesson, students work collaboratively to brainstorm questions, design procedures for testing their predictions, carry out investigations, and ask thoughtful questions about other students' conclusions. This mirrors the social context in which "real science" takes place.

What Makes a Good Question?

Questioning lies at the heart of inquiry and is a habit of mind that should be encouraged in any learning setting. According to *Inquiry and the National Science Education Standards* (NRC 2000, p. 24–25), "Fruitful inquiries evolve from questions that are meaningful and relevant to students, but they also must be able to be answered by students' observations and scientific knowledge they obtain from reliable sources." In an inquiry classroom, the teacher plays an important role in helping students identify questions that can lead to interesting and productive investigations, questions that are accessible, manageable, and appropriate to students' developmental level.

Inquiry With Interesting Objects

A great way to encourage questioning is to provide students with interesting objects to stimulate their curiosity. As stated by Doris Ash, "Curiosity drives the inquiry process—it generates questions and a search for answers." (1999, p. 754) You can use fun, interesting, or unusual objects that you have collected, but you can often pique student curiosity by simply giving students ordinary objects and challenging them to look at the objects in a scientific way. A helpful tool that you can provide for this activity is an O-W-L (**O**bservations, **W**onderings, **L**earnings) chart. In the first column of the chart, students write their observations of the object and in the second column, they write their wonderings. These wonderings can propel students into meaningful inquiry experiences. Objects useful for this activity include

Observing interesting objects

Writing a question for a question sort

pumpkins, leaves, seashells, plant galls, seeds, potatoes, pinecones, jumping beans, plastic animal models, inexpensive plastic toys, and fossils. After students have had a chance to carefully observe, to wonder, and to share their questions with others, you can use the question-sorting activity that follows as a springboard to inquiry investigations about the objects.

Question Sort

One of the most important skills students can develop in science is to understand which ques-

tions can be answered by investigation and which cannot. The teacher plays a critical role in guiding the kinds of questions the students pose. Students often ask *why* questions, which cannot be addressed by scientific investigations. For example, "Why does gravity make things fall toward Earth?" is a question that would be impossible to answer in the school setting.

Testable questions, on the other hand, generally begin with *how can, does, what if,* or *which* and can be investigated using controlled procedures. For example, encouraging students to ask questions such as "How can you slow the fall of an object?" "Which object falls faster, a marble or a basketball?" or "What materials work best for constructing a toy parachute?" guides them toward investigations that can be done in the classroom.

One way you can help students learn which types of questions are testable and which are not testable is to do a "question sort." In this activity, you start by providing students with a common experience on a science topic using thought-provoking objects, readings, video clips, and so on. Next, you ask students to write a question about the topic on a sentence strip or sticky note. Collect all of the sentence strips and read the questions aloud to the class. Explain that the type of investigation a scientist does depends on the

Sample Question Sort	
Research Questions	**Testable Questions**
Why are pumpkins orange?	Are there more lines on bigger pumpkins than on smaller ones?
How big was the biggest pumpkin ever grown?	Do larger pumpkins have larger seeds than smaller pumpkins?
Where do pumpkins grow?	Do larger pumpkins have more seeds than smaller pumpkins?
How do you make pumpkin pie?	How much less does a pumpkin weigh after it has been carved?
	What happens to a carved pumpkin after one month? after two months?

Results of a question sort

questions he or she asks. As a group, you can sort the students' questions into "researchable questions" that can be answered using reliable sources of scientific information, and "testable questions" that can be answered by observing, measuring, or doing an experiment.

After taking part in a question sort, students will begin to realize that with a little tweaking, some researchable questions can be turned into testable questions. Young students may have difficulty coming up with testable questions. In this case, it is appropriate that you come up with the questions. From your examples, students will gradually begin to learn what testable questions look and sound like.

The next step is to have the class select one of the testable questions and discuss ways to investigate the question. For example, the question "Do larger pumpkins have more seeds than smaller pumpkins?" could be answered by cutting open several large pumpkins and several small pumpkins and comparing the number of seeds inside each. After investigating the question, students can brainstorm ways to communicate their results, such as with pictures, data tables, graphs, and poster presentations.

Helping students select developmentally appropriate questions is also important. For example, "What will the surface of the Moon look like in a hundred years?" is a question that is scientific but much too complex for elementary students to investigate. A more developmentally appropriate question might be "How does the size of a meteorite affect the size of the crater it makes?" This question can be tested by dropping different-sized marbles into a pan of sand, simulating how meteors hit the Moon's surface. It is essential to help students formulate age-appropriate and testable questions to ensure that their investigations are both engaging and productive.

The Role of the Teacher

Teaching science through inquiry requires that the teacher take on a different role than the traditional science teacher. "In the inquiry classroom, the teacher's role becomes less involved with direct teaching and more involved with modeling, guiding, facilitating, and continually assessing student work." (Ashand Kluger-Bell 1999, p. 82) One way to guide students and assess their progress as they are engaged in inquiry processes is to ask thoughtful probing questions. Some suggested questions to ask students while they are involved in inquiry follow:

- What would happen if you …?
- What might you try instead?
- What does this remind you of?
- What can you do next time?
- What do you call the things you are using?
- How are you going to do that?
- Is there anything else you could use or do?
- Why did you decide to try that?
- Why do you think that will work?
- Where could you get more information?
- How do you know?
- What is your evidence?

Variations Within Classroom Inquiry

Inquiry-based teaching can vary widely in the amount of guidance and structure you choose to provide. Table 3.2 describes these variations for each of the five essential features of inquiry.

The most open form of inquiry takes place in the variations on the right-hand column of the Inquiry Continuum. Most often, students do not have the abilities to begin at that point. For example, students must first learn what makes a question scientifically oriented and testable before they can begin posing such questions themselves. The extent to which you structure what students do determines whether the inquiry is *guided* or

open inquiry. The more responsibility you take, the more guided the inquiry. The more responsibility the students have, the more open the inquiry. Guided inquiry experiences, such as those on the left-hand side of the inquiry continuum, can be effective in focusing learning on the development of particular science concepts. Students, however, must have open inquiry experiences, such as those in the right column of the Inquiry Continuum, to develop the fundamental abilities necessary to do scientific inquiry. "Inquiry investigations in the classroom can be highly structured by the teachers so that students proceed toward known outcomes, or inquiry investigations can be free-ranging explorations of unexplained phenomena. Both have their place in science classrooms." (NRC 2000).

One common misconception about inquiry is that all science subject matter should be taught through inquiry. It is not possible or practical to teach all science subject matter through inquiry (NRC 2000). For example, you would not want to teach lab safety through inquiry. Good science teaching requires a variety of approaches and models. *More Picture-Perfect Science Lessons* combines a guided-inquiry investigation with an open-inquiry investigation. Dunkhase (2000) refers to this approach as "coupled inquiry." In *More Picture-Perfect Science Lessons*, the guided inquiries are the lessons presented in each chapter. The lessons generally fall on the left-hand (teacher-guided) side of the Inquiry Continuum. The "Inquiry Place" suggestion box (discussed in depth later in this chapter) at the end of each lesson will produce experiences falling more toward the right-hand, or learner self-directed, side of the Inquiry Continuum.

Inquiry Place

As we mentioned earlier, a box called *Inquiry Place* is provided at the end of each lesson to help you move your students toward more open inquiries. The Inquiry Place lists questions related to the lesson that students may select to investigate. Students may also use the questions as examples

Table 3.2 Inquiry Continuum

Teacher Guided ←————→ Learner Self-Directed				
ESSENTIAL FEATURE	**VARIATIONS**			
1 Learners are engaged in scientifically oriented questions.	Learner engages in question provided by teacher or materials	Learner sharpens or clarifies the question provided	Learner selects among questions, poses new questions	Learner poses a question
2 Learners give priority to evidence, which allows them to develop and evaluate explanations that address scientifically oriented questions.	Learner given data and told how to analyze	Learner given data and asked how to analyze	Learner directed to collect certain data	Learner determines what constitutes evidence and collects it
3 Learners formulate explanations from evidence to address scientifically oriented questions.	Learner provided with evidence	Learner given possible ways to use evidence to formulate explanations	Learner guided in process of formulating explanations from evidence	Learner formulates explanation after summarizing evidence
4 Learners evaluate their explanations in light of alternative explanations, particularly those reflecting scientific understanding.	Learner told connections	Learner given possible connections	Learner directed toward areas and sources of scientific knowledge	Learner independently examines other resources and forms the links to explanations
5 Learners communicate and justify their proposed explanations.	Learner given steps and procedures for communication	Learner provided broad guidelines to sharpen communication	Learner coached in development of communication	Learner communicates and justifies explanations
Teacher Guided ←————→ Learner Self-Directed				

Adapted from *Inquiry and the National Science Education Standards: A Guide for Teaching and Learning* (NRC 2000).

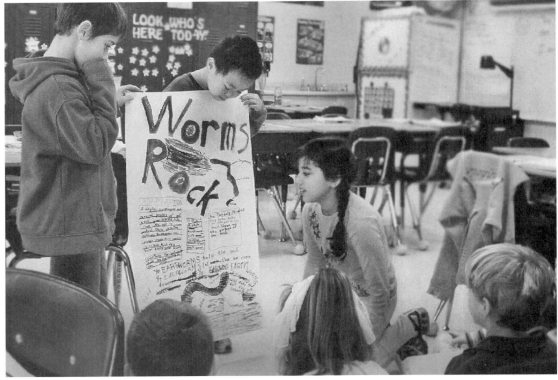

Presenting at a poster session

to help them generate their own scientifically oriented and testable questions. After selecting one of the questions in the box or formulating their own questions, students can make predictions, design investigations to test their predictions, collect evidence, devise explanations, examine related resources, and communicate their findings.

The Inquiry Place boxes suggest that students share the results of their investigations with each other through a poster session. Scientists, engineers, and researchers routinely hold poster sessions to communicate their findings. Here are some suggestions for poster sessions:

- Posters should include a title, the researchers' names, a brief description of the investigation, and a summary of the main findings.

- Observations, data tables, and/or graphs should be included as evidence to justify conclusions.

- The printing should be large enough that people can read it from a distance.

- Students should have the opportunity to present their posters to the class.

- The audience in a poster session should examine the evidence, ask thoughtful questions, identify faulty reasoning, and suggest alternative explanations to presenters in a polite, respectful manner.

Not only do poster sessions mirror the work of real scientists, but they also provide you with excellent opportunities for authentic assessment. Another way to share students' posters is a gallery walk. In a gallery walk, students put their posters on display for their classmates to view and critique. Students taking the gallery walk use sticky notes to post suggestions, questions, and praise directly onto their classmates' posters. Writing on sticky notes encourages interaction, and the comments provide immediate feedback for the "exhibitors." Here are some guidelines for a gallery walk:

- All necessary information about the investigation should be included on the poster

A gallery walk

because students will not be giving an oral presentation.

- Like a visit to an art gallery, the gallery walk should be done quietly. Students should be respectful of their classmates' poster displays.

- Students should have the opportunity to read the comments about their own posters and make changes if necessary.

Implementing the guided inquiries in this book along with the Inquiry Place suggestions at the end of each lesson provides a framework for moving from teacher-guided to learner self-directed inquiry. The Inquiry Place Think Sheet on page 28 (Table 3.3) can help students organize their own inquiries.

An example of how the Inquiry Place can be used to give students the opportunity to engage in an open inquiry follows. This particular example is from "That Magnetic Dog," Chapter 13. In that lesson, students discover through exploration that not all metals are magnetic. This is how one teacher chose to use the Inquiry Place following that guided inquiry lesson:

After completing the magnetic dog lesson, Mrs. Quill begins a discussion about magnets. She and her students talk about their experiences with magnets.

Mrs. Quill: We have a lot of different-sized magnets in the classroom. I wonder if the strength of a magnet has to do with how big it is. For example, are bigger magnets stronger than smaller magnets? How can we use these supplies to find the answer to the question?

Mrs. Quill writes the question on the board: "Are bigger magnets stronger than smaller magnets?"

Pedro: We can use a big magnet and a small magnet and count how many paper clips each one picks up.

Mrs. Quill: Should we just use two magnets?

Keira: No, I think we should use lots of magnets to make sure our answer is right.

Mrs. Quill: Great idea! Now, how can we make sure the experiment is fair? [Silence]

Mrs. Quill: What things should we keep the same? For example, can we use different sized paper clips?

Kevin: No, that would not be fair. We should use the same sized paper clips with all the magnets.

Carlos: And we should make sure we put the paper clips on the magnets the same way each time. Maybe we should make a hook out of one paper clip and put all the other paper clips on the hook. If we do it the same way for all the magnets, the experiment will be fair.

Mrs. Quill: Sounds great! Now, before we begin we need to think about how we will record our data.

Eva: We can make a table with "Size of Magnet" on one side and "Number of Paper Clips" on the other side.

Mrs. Quill: Great idea! Let's make a data table on the board. I think we are ready to begin the experiment. Let's make some predictions first.

Mrs. Quill and her class make predictions and then perform the experiment together. They record their data on the board and use it as evidence to answer the question. After some discussion, the students reach the conclusion that the size of a magnet does not determine its strength.

Mrs. Quill: Now that we have answered my question, I wonder if you have any questions about magnets that we could investigate with our lab materials?

Mrs. Quill passes out the Inquiry Place Think Sheet to each student. After providing students time to brainstorm some questions about magnets, Mrs. Quill asks them to share their favorite ones.

Keira: What are magnets made of?

Mrs. Quill: That's a good question, Keira. Can we use our lab materials to find the answer?

Keira: No.

Mrs. Quill: How could we find the answer to that question?

Keira: Maybe at the library or on the internet.

Mrs. Quill: Yes. Maybe we can look for that answer next time we are in the library or computer lab. Let's try to think of questions that we can answer using the materials we have.

Yushi: We can find out if magnetic force can go through plastic.

Joshua: Or, can a magnet work through water?

Keira: What happens when two magnets touch?

Mrs. Quill writes the questions on the board.

Mrs. Quill: Excellent! These three questions can be answered by investigating with our lab materials. Choose one of the questions that you would like to investigate and write it down for number 2.

Mrs. Quill provides time for students to think about which question they want to investigate and to write it down. She then forms teams of students who have chosen the same question.

> *Mrs. Quill: Now that you have formed your teams, complete the rest of the Inquiry Place Think Sheet together. Signal when your experiment is planned and you are ready for the teacher checkpoint.*

Mrs. Quill circulates to ask questions and check progress as teams complete the Inquiry Place Think Sheet. Students finish planning the investigations and look forward to completing them the next day. They will share their findings during a poster session later in the week.

Picture Books That Support Inquiry

We are always on the lookout for books that support the inquiry process and scientific habits of mind. Below are some of our favorites. Reading some of these books aloud can spark conversations with your students about the nature of science.

Bender, L. 2007. *Magnification: A closer look*. Minneapolis: Picture Window Books.

Bender, L. 2007. *Science safety: Being careful*. Minneapolis: Picture Window Books.

Dotlich, R. K. 2006. *What is science?* New York: Henry Holt and Company.

Eboch, C. 2007. *Science measurements: How heavy? How long? How hot?* Minneapolis: Picture Window Books.

Eboch, C. 2007. *Science tools: Using machines and instruments*. Minneapolis: Picture Window Books.

Lehn, B. 1999. *What is a scientist?* Brookfield, CT: Millbrook Press.

Thinking Like a Scientist Series. 1999. New York: Newbridge Educational Publishing. Titles include *Let's Experiment, Science Tools, Exploring Everyday Wonders, Being a Scientist,* and *A Closer Look.*

References

Ash, D. 1999. The process skills of inquiry, In *Foundations Volume 2: Inquiry—thoughts, views, and strategies for the K–5 classroom*. Arlington, VA: Division of Elementary, Secondary, and Informal Education in conjunction with the Division of Research, Evaluation, and Communication, National Science Foundation.

Ash, D., and B. Kluger-Bell. 1999. Identifying inquiry in the K–5 classroom. In *Foundations Volume 2: Inquiry—thoughts, views, and strategies*. Arlington, VA: Division of Elementary, Secondary, and Informal Education in conjunction with the Division of Research, Evaluation, and Communication, National Science Foundation.

Bransford, J. D., A. L. Brown, and R. Cocking, eds. 1999. *How people learn: Brain, mind, experience, and school*. Washington, DC: National Academy Press.

Dunkhase, J. 2000. *Coupled inquiry: An effective strategy for student investigations*. Des Moines, IA: Paper presented at the Iowa Science Teachers Section Conference.

Dyasi, H. 1999. What children gain by learning through inquiry. In *Foundations Volume 2: Inquiry—thoughts, views, and strategies for the K–5 classroom*. Arlington, VA: Division of Elementary, Secondary, and Informal Education in conjunction with the Division of Research, Evaluation, and Communication, National Science Foundation.

National Research Council (NRC). 1996. *National science education standards*. Washington, DC: National Academy Press. Available online at *books.nap.edu/books/0309053269/html/index.html*.

National Research Council (NRC). 2000. *Inquiry and the National Science Education Standards: A guide for teaching and learning*. Washington, DC: National Academy Press. Available online at *www.nap.edu/books/0309064767/html*.

National Science Foundation (NSF). 1999. *Inquiry: Thoughts, views, and strategies for the K–5 classroom: Foundations*, Vol. 2. Arlington, VA: Division of Elementary, Secondary, and Information Education. National Science Foundation.

Wandersee, J. H., J. J. Mintzes, and J. D. Novak. 1994. Research on alternative conceptions in science. In *Handbook of research on science teaching and learning*, ed. D. L. Gable, 177–210. New York: Macmillan.

Table 3.3 **Inquiry Place Think Sheet**

Name_____

Inquiry Place Think Sheet

1 Topic: _____

2 My questions about: _____

_____

_____

3 My testable question: _____

_____

4 My prediction: _____

_____

5 Steps I will follow to investigate my question:

_____

_____

_____

_____

6 Materials I will need: _____

_____

7 How I will share my findings: _____

_____

_____

_____

Checkpoint

☐

NATIONAL SCIENCE TEACHERS ASSOCIATION

BSCS 5E Instructional Model

The guided inquiries in this book are designed using the BSCS 5E Instructional Model, commonly referred to as the 5E model (or the 5Es). Developed by the Biological Sciences Curriculum Study (BSCS), the 5E model is a learning cycle based on a constructivist view of learning. Constructivism embraces the idea that learners bring with them preconceived ideas about how the world works. According to the constructivist view, "learners test new ideas against that which they already believe to be true. If the new ideas seem to fit in with their pictures of the world, they have little difficulty learning the ideas … if the new ideas don't seem to fit the learners' picture of reality then they won't seem to make sense. Learners may dismiss them … or eventually accommodate the new ideas and change the way they understand the world" (Colburn 2003, p. 59). The objective of a constructivist model, therefore, is to provide students with experiences that make them reconsider their conceptions. Then, students "redefine, reorganize, elaborate, and change their initial concepts through self-reflection and inter-action with their peers and their environment" (Bybee 1997, p. 176). The 5E model provides a planned sequence of instruction that places students at the center of their learning experiences, encouraging them to explore, construct their own understanding of scientific concepts, and relate those understandings to other concepts. An explanation of each phase of the BSCS 5E model—*engage, explore, explain, elaborate,* and *evaluate*—follows.

engage

The purpose of this introductory stage, *engage*, is to capture students' interest. Here you can uncover what students know and think about a topic as well as determine their misconceptions. Engagement activities might include a reading, a demonstration, or other activity that piques students' curiosity.

Engage: Mrs. Custis generates curiosity about coral reefs by inviting students to interact with the text during a read-aloud ("Over in the Ocean," Chapter 11).

explore

In the *explore* stage, you provide students with cooperative exploration activities, giving them common, concrete experiences that help them begin constructing concepts and developing skills. Students can build models, collect data, make and test predictions, or form new predictions. The purpose is to provide hands-on experiences you can use later to formally introduce a concept, process, or skill.

Explore: Mrs. Quill challenges her students to design a roller coaster ("Roller Coasters," Chapter 14).

explain

In the *explain* stage, learners articulate their ideas in their own words and listen critically to one another. You clarify their concepts, correct misconceptions, and introduce scientific terminology. It is important that you clearly connect the students' explanations to experiences they had in the *engage* and *explore* phases.

Explain: Students in Miss Calpin's class create posters to help them explain which objects were attracted to a magnet ("That Magnetic Dog," Chapter 13).

elaborate

At the *elaborate* point in the model, some students may still have misconceptions, or they may understand the concepts only in the context of the previous exploration. Elaboration activities can help students correct their remaining misconceptions and generalize the concepts in a broader context. These activities also challenge students to apply, extend, or elaborate upon concepts and skills in a new situation, resulting in deeper understanding.

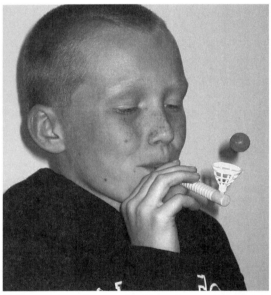

Elaborate: A student applies his understanding of risks and benefits by testing toys ("Imaginative Inventions," Chapter 19).

evaluate

In the *evaluate* phase, you evaluate students' understanding of concepts and their proficiency with various skills. You can use a variety of formal and informal procedures to assess conceptual understanding and progress toward learning outcomes. The evaluation phase also provides an opportunity for students to test their own understanding and skills.

Although the fifth phase is devoted to evaluation, a skillful teacher evaluates throughout the 5E model, continually checking to see if students need more time or instruction to learn the key points in a lesson. Ways to do this include informal questioning, teacher checkpoints, and class discussions. Each lesson in *More Picture-Perfect Science Lessons* also includes a formal evaluation such as a written quiz or poster session. These formal evaluations take place at the end of the lesson. A good resource for more information and practical suggestions for evaluating student understanding throughout the 5Es is *Seamless Assessment in Science: A Guide for Elementary and Middle School Teachers* by Abell and Volkmann (2006).

Evaluate: A student shows what he has learned about sun safety.

Cycle of Learning

The 5Es are listed above in linear order—engage, explore, explain, elaborate, and evaluate—but the model is most effective when you use it as a cycle of learning as in Figure 4.1 (p. 32).

Each lesson begins with an engagement activity, but students can reenter the 5E model at other points in the cycle. For example, in "Wiggling Worms," Chapter 10, students *explore* the characteristics of earthworms. Then they *explain* their earthworm observations and compare them to their classmates'. Next, the students reenter the *explore* phase by performing an experiment to find out if worms prefer damp or dry places. Moving from the *explain* phase back into the *explore* phase gives students the opportunity to add to the knowledge they have constructed so far in the lesson by participating in additional hands-on explorations.

The traditional roles of the teacher and student are virtually reversed in the 5E model. Students take on much of the responsibility for learning as they construct knowledge through discovery, whereas in traditional models the teacher is responsible for dispensing information to be learned by the students. Table 4.1 shows actions of the teacher that are consistent with the 5E model and actions that are inconsistent with the model.

In the 5E model, the teacher acts as a guide: raising questions, providing opportunities for exploration, asking for evidence to support student explanations, referring students to existing explanations, correcting misconceptions, and coaching students as they apply new concepts. This model differs greatly from the traditional format of lecturing, leading students step-by-step to a solution, providing definite answers, and testing isolated facts. The 5E model requires the students to take on much of the responsibility for their own learning. Table 4.2 (p. 33) shows the actions of the student that are consistent with the 5E model and those that are inconsistent with the model.

Using Children's Picture Books in the 5Es

Both fiction and nonfiction picture books can be valuable components of the 5E model when placed strategically within the cycle. We often begin lessons with a fiction book to pique students' curiosity or motivate them to want to learn more about a science concept. For example, Chapter 6 ("Bubbles") begins with a story about a young boy

Table 4.1 **The BSCS 5Es Teacher**

	What the teacher does	
	CONSISTENT with the BSCS 5E model	*INCONSISTENT* with the BSCS 5E model
engage	• Generates interest and curiosity • Raises questions • Assesses current knowledge, including misconceptions	• Explains concepts • Provides definitions and conclusions • Lectures
explore	• Provides time for students to work together • Observes and listens to students as they interact • Asks probing questions to redirect students' investigations when necessary	• Explains how to work through the problem or provides answers • Tells students they are wrong • Gives information or facts that solve the problem
explain	• Asks for evidence and clarification from student • Uses students' previous experiences as a basis for explaining concepts • Encourages students to explain concepts and definitions in their own words, then provides scientific explanations and vocabulary	• Does not solicit the students' explanations • Accepts explanations that have no justification • Introduces unrelated concepts or skills
elaborate	• Expects students to apply scientific concepts, skills, and vocabulary to new situations • Reminds students of alternative explanations • Refers students to alternative explanations	• Provides definite answers • Leads students to step-by-step solutions to new problems • Lectures
evaluate	• Observes and assesses students as they apply new concepts and skills • Allows students to assess their own learning and group process skills • Asks open-ended questions	• Tests vocabulary words and isolated facts • Introduces new ideas or concepts • Promotes open-ended discussion unrelated to the concept

Adapted from *Achieving Scientific Literacy: From Purposes to Practices* (Bybee 1997).

Table 4.2 **The BSCS 5Es Student**

	What the student does	
	CONSISTENT with the BSCS 5E model	**INCONSISTENT** with the BSCS 5E model
engage	● Asks questions such as, "Why did this happen? What do I already know about this? What can I find out about this?" ● Shows interest in the topic	● Asks for the "right" answer ● Offers the "right" answer ● Insists on answers and explanations
explore	● Thinks creatively, but within the limits of the activity ● Tests predictions and hypotheses ● Records observations and ideas	● Passively allows others to do the thinking and exploring ● "Plays around" indiscriminately with no goal in mind ● Stops with one solution
explain	● Explains possible solutions to others ● Listens critically to explanations of other students and the teacher ● Uses recorded observations in explanations	● Proposes explanations from "thin air" with no relationship to previous experiences ● Brings up irrelevant experiences and examples ● Accepts explanations without justification
elaborate	● Applies new labels, definitions, explanations, and skills in new but similar situations ● Uses previous information to ask questions, propose solutions, make decisions, design experiments ● Records observations and explanations	● "Plays around" with no goal in mind ● Ignores previous information or evidence ● Neglects to record data
evaluate	● Demonstrates an understanding of the concept or skill ● Answers open-ended questions by using observations, evidence, and previously accepted explanations ● Evaluates his/her own progress and knowledge	● Draws conclusions, not using evidence or previously accepted explanations ● Offers only yes-or-no answers and memorized definitions or explanations ● Fails to express satisfactory explanations in his/her own words
Adapted from *Achieving Scientific Literacy: From Purposes to Practices* (Bybee 1997).		

who uses a magic bubble maker to blow bubbles in all kinds of shapes … a kangaroo, a snake, and a cat, and so on. This read aloud during the engage phase sets up the question, "Are all free-floating bubbles round?" and is followed by an investigation to find out. A storybook, however, might not be appropriate to use during the explore phase of the 5Es in which you want students to take part in hands-on concrete experiences.

You should also avoid using books too early in the learning cycle that contain a lot of scientific terminology or "give away" information students could discover on their own. It is important for students to have opportunities to construct meaning and articulate ideas in their own words before being introduced to scientific vocabulary. Nonfiction books, therefore, are most appropriate to use in the explain phase only after students have had these opportunities. For example, in the explain phase of Chapter 13 ("That Magnetic Dog"), students first develop and share their own hypotheses about why certain objects are attracted to magnets. Then, students compare their hypotheses to the information presented in the nonfiction book *Magnetic and Nonmagnetic*. They are also introduced to the terms *magnetic* and *nonmagnetic* at this point.

Thoughtful placement of fiction and nonfiction picture books within the BSCS 5E Instructional Model can motivate students to learn about science, allow them to evaluate their findings in the light of alternative explanations, and help them understand scientific concepts and vocabulary without taking away from the joy of discovery.

References

Abell, S. K., and M.J. Volkmann. 2006. *Seamless assessment in science: A guide for elementary and middle school teachers.* Chicago, IL: Heinemann and Arlington, VA: NSTA Press.

Bybee, R. W. 1997. *Achieving scientific literacy: From purposes to practices.* Portsmouth, NH: Heinemann.

Colburn, A. 2003. *The lingo of learning: 88 education terms every science teacher should know.* Arlington, VA: NSTA Press.

National Science Education Standards

The National Science Education Standards were published in 1996 with the intent of helping our nation's students achieve science literacy. They outline what students at different grade ranges need to know, understand, and be able to do to be considered scientifically literate. The book, *National Science Education Standards*, defines scientific literacy as "the knowledge and understanding of scientific concepts and processes required for personal decision making, participation in civic and cultural affairs, and economic productivity" (NRC 1996, p. 22). The Standards are designed to be achievable by all students, regardless of their backgrounds or characteristics, and are based on the premise that science is an active process. "Learning science is something that students do, not something that is done to them. 'Hands-on' activities, while essential, are not enough. Students must have 'minds-on' experiences as well" (NRC 1996, p. 20). *More Picture-Perfect Science Lessons* embraces this philosophy by engaging students in meaningful, hands-on science experiences that require them to construct their own knowledge.

The lesson objectives in *More Picture-Perfect Science Lessons* have been adapted from the Standards. Because the content Standards themselves are not very specific, we consulted the bulleted lists of fundamental concepts and principles that underlie each Standard when developing lesson objectives. We refer to these as "fundamental understandings" and have included them at the beginning of each lesson in the box titled "Lesson Objectives: Connecting to the Standards." For example, in "Bubbles," Chapter 6, one of the content Standards for K–4 addressed in the lesson is "Content Standard A: Science as Inquiry." The fundamental understanding from K–4 Content Standard A that we selected for this particular lesson is "Understand that 'scientific investigations involve asking and answering a question and comparing the answer to what scientists already know about the world'" (NRC 1996, p. 123).

The fundamental understandings used to develop the lesson objectives in *More Picture-Perfect Science Lessons* were adapted from the "Guide to the Content Standards" sections of the *National Science Education Standards*, which describes the fundamental ideas that underlie the Standards.

Because *More Picture-Perfect Science Lessons* focuses on student learning of specific science content objectives in kindergarten through grade four, this chapter outlines only the National Science Education Content Standards for that grade band. These content Standards are listed in Table 5.1 (p. 36).

Table 5.2 (p. 37) shows the correlation between the lessons presented in this book and the National Science Education Standards for grades K–4.

Reference

National Research Council. 1996. *National Science Education Standards*. Washington, DC: National Academy Press.

Table 5.1 **National Science Content Standards K–4**

Unifying Concepts and Processes
- Systems, order, and organization
- Evidence, models, and explanation
- Change, constancy, and measurement
- Evolution and equilibrium
- Form and function

Content Standard A: Science as Inquiry
- Abilities necessary to do science inquiry
- Understandings about science inquiry

Content Standard B: Physical Science
- Properties of objects and materials
- Position and motion of objects
- Light, heat, electricity, and magnetism

Content Standard C: Life Science
- The characteristics of organisms
- Life cycles of organisms
- Organisms and environments

Content Standard D: Earth and Space Science
- Properties of Earth materials
- Objects in the sky
- Changes in Earth and sky

Content Standard E: Science and Technology
- Abilities of technological design
- Understandings about science and technology
- Abilities to distinguish between natural objects
  and objects made by humans

Content Standard F: Science in Personal and Social Perspectives
- Personal health
- Characteristics and changes in populations
- Types of resources
- Changes in environments
- Science and technology in local challenges

Content Standard G: History and Nature of Science
- Science as a human endeavor

National Science Education Standards (NRC 1996).

Table 5.2 Connecting to the National Science Education Standards

	Content Standard A Science as Inquiry	Content Standard B Physical Science	Content Standard C Life Science	Content Standard D Earth and Space Science	Content Standard E Science and Technology	Content Standard F Science in Personal and Social Perspectives	Content Standard G History and Nature of Science
Chapter 6: Bubbles	X						
Chapter 7: How Big Is a Foot?					X		X
Chapter 8: Hear Your Heart	X					X	
Chapter 9: Loco Beans	X		X				
Chapter 10: Wiggling Worms	X		X				
Chapter 11: Over in the Ocean	X		X				
Chapter 12: Be a Friend to Trees	X		X			X	
Chapter 13: That Magnetic Dog	X	X					
Chapter 14: Roller Coasters	X	X					
Chapter 15: Mirror, Mirror	X	X					
Chapter 16: If You Find a Rock	X			X			
Chapter 17: Sunshine on My Shoulders	X			X			
Chapter 18: Stargazers	X			X			
Chapter 19: Imaginative Inventions					X		
Chapter 20: A Sense of Wonder			X				X

Bubbles

Description

Learners engage in a scientific investigation to answer the question, "Are free-floating bubbles always round?" By experimenting with different-shaped bubble wands and then reading a nonfiction book to support their findings, learners collect evidence to answer the question and then share their findings with others by creating a poster.

Suggested Grade Levels: K–2

Lesson Objectives Connecting to the Standards

Content Standard A: Science as Inquiry, Abilities Necessary to Do Scientific Inquiry

- Ask a question about objects, organisms, and events in the environment.
- Design and conduct simple experiments to answer questions.
- Use data to construct reasonable explanations.

Content Standard A: Science as Inquiry, Understandings About Scientific Inquiry

- Understand that scientific investigations involve asking and answering a question and comparing the answer to what scientists already know about the world.
- Realize that good explanations are based on evidence from investigations.
- Understand that scientists make the results of their investigations public.

Featured Picture Books

Title	*Bubble, Bubble*	*Pop! A Book About Bubbles*
Author	Mercer Mayer	Kimberly Brubaker Bradley
Illustrator	Mercer Mayer	Margaret Miller
Publisher	Gingham Dog Press	HarperTrophy
Year	1973	2001
Genre	Story	Nonfiction Information
Summary	With magic bubble solution, a boy discovers that he can blow any kind of bubble imaginable: a kangaroo, a bird, a car, or a boat.	Simple text explains how soap bubbles are made, why floating bubbles are always round, and what makes them pop.

Time Needed

This lesson will take several class periods. Suggested scheduling is as follows:

Day 1: **Engage** with *Bubble, Bubble* read aloud and **Explore** with Bubble Shapes Investigation.

Day 2: **Explain** with Looking at the Data and *Pop! A Book about Bubbles* read aloud.

Day 3: **Elaborate** with Bubble Toy Testing and **Evaluate** with Sharing Our Findings.

Materials

Bubble solution (see recipe)

Small cups (1 per pair of students)

Round bubble wands (1 per pair of students)

Pipe cleaners (3 per pair of students)

Assortment of bubble toys

> **Recipe for Bubble Solution**
>
> (From *Pop! A Book About Bubbles*)
>
> 1 tbsp. of dishwashing liquid
>
> 1 tbsp. of corn syrup or glycerin
>
> 9 tbsp. of water.
>
> Stir very gently, so you don't make foam.

Student Pages

Bubble Shapes Data Table

Background

Children can experience the nature of scientific investigations at an early age. The basic ideas about scientific investigations are actually quite simple. Scientists ask questions, design investigations to answer the questions, collect data, read or talk to other experts to find out what is already known, come up with answers based on the data, and share their results with others. This lesson is not just about bubbles. It is also about the nature of scientific investigations—the notion that people are more likely to accept your ideas if you can give good reasons for them.

Students will discover that no matter the shape of the bubble wand, bubbles always take on a round (spherical) shape after they break free of the wand. Free-floating bubbles are round because they always hold the gas inside of them with the least possible surface area. If an elongated bubble is formed, it will eventually form a sphere or break into two spheres before it bursts. Note that bubbles would be *perfect* spheres only under ideal conditions of no gravity, no air movement, and no air resistance, but students will be able to observe a basically round shape once a bubble becomes free floating.

engage

Bubble, Bubble Read Aloud

Making Connections: Text to Self

Show students the book *Bubble, Bubble* and ask

? Have you ever blown a bubble? What did you use? (Bubble solution and a wand, chewing gum, and dish soap are among likely answers)

Read the book aloud to students, then ask

? What were some of the bubble shapes we saw in the book? (Answers might include: a boat, a snake, and an elephant.)

? Can you really blow bubbles in the shape of a boat, a snake, or an elephant?

explore

Bubble Shapes Investigation

Take students outside, and give each pair a cup of bubble solution and a round bubble wand. Allow time for guided discovery of bubble shapes. As students are blowing bubbles ask

? What can you observe about the shapes of the bubbles? (Students may notice that when a bubble is still attached to the wand, it can be elongated or irregular in shape. But when it breaks free of the wand, it is round.)

? What else can you observe about the bubbles? (Students may observe the colors in the bubbles, they splatter when they pop, they move around with the wind, and they can be different sizes.)

SAFETY

Caution students not to get the solution in their eyes or mouths.

Testing a round bubble wand

Explain that when bubbles break free from the wand, we say that they are *free floating*. Ask

? Why do you think the free-floating bubbles you observed were always round? (Students may say that it is because the wands they used are round.)

? Are free-floating bubbles always round? How could we find out?

Allow students time to come up with ideas about how they might design an experiment to find the answer to the question. Then suggest that one way this question can be answered is to try blowing bubbles with different-shaped bubble wands. Give each student a copy of the Bubble Shapes Data Table, and explain that when scientists want to know the answer to a question, they design an *experiment*. When they do the experiment, they write down what they learn from it in an organized way. Scientists call this information *data*. Explain that in the first column of the data table, the shapes of the wands are recorded, and, in the second column of the data table, the shape of the bubble is recorded. Give each pair of students three pipe cleaners and have them bend the pipe cleaners into a round wand, a square wand, and a triangular wand, as shown on the data table.

SAFETY
Caution students not to get the solution in their eyes or mouths.

When students have finished making the wands and before they begin blowing bubbles, have them make a prediction. Ask

? When you use the round wand, what shape do you think the free-floating bubble will be?

? When you use the square wand, what shape do you think the free-floating bubble will be?

? When you use the triangular wand, what shape

do you think the free-floating bubble will be?

Then give each pair a cup of bubble solution. Students can then try blowing bubbles with the different shaped wands and record the shape of the free-floating bubbles in their data tables.

explain

Looking at the Data and *Pop! A Book About Bubbles* Read Aloud

Have students compare their data tables and discuss their results. Students will realize that no matter what the shape of the wand is, the free-floating bubbles are always round.

Determining Importance

Tell students that one way scientists find answers to

When bubbles break free from the wand, they become round.

their questions is to do experiments. They also read nonfiction books to find answers and to check how the results of their experiments compare to what is already known. Show students the cover of *Pop! A Book About Bubbles,* and ask them to signal when they hear anything about whether or not free-floating bubbles are always round. After reading the book aloud, discuss how the information from the reading ("Bubbles are always round.") compares to their results from the bubble-shapes investigation.

elaborate
Bubble Toy Testing

Tell students that in science, the more evidence you have, the more sure you can be of your answer. To gather more evidence that all free-floating bubbles are round, provide students with various bubble toys. Then have them predict the shape of the free-floating bubbles that can be made with each toy. Give them time to try out the toys and observe the shapes of the bubbles. Next, have them describe what they observed about the bubble shapes. Students should be able to

> **SAFETY** Caution students not to get the solution in their eyes or mouths.

explain that, no matter what toy they try, the bubbles are always round once they become free-floating.

evaluate
Sharing Our Findings

Explain that when scientists find the answers to the questions they are wondering about, they often share the answers with other scientists. Sometimes, they do this through creating a poster. On the poster, a scientist might list the following things:

1 The question he or she is wondering about.

2 The answer to the question.

3 The evidence that supports the answer.

Create a class poster to display in the classroom or hallway that contains the information above. As you create the poster together ask

? What was our question? (Are free-floating bubbles always round?)

? What is our answer? (Yes, all free-floating bubbles are round.)

? What is our evidence? (The information on the data table. The information in *Pop! A Book about Bubbles.* The observations we made when we tested the bubble toys.)

Inquiry Place

Have students brainstorm testable questions such as

? Can you blow a bubble with plain water? milk? juice?

? How does the speed of blowing affect the size of the bubble?

? How does the size of the wand affect the size of the bubble?

Students can choose a question to investigate in teams or as a class. After they make predictions, have them conduct an experiment to test their predictions. Students can present their findings in a poster session or gallery walk.

More Books to Read

Hulme J. 1999. *Bubble trouble.* Danbury, CT: Children's Press.

> Summary: Rhyming verse and colorful, round-headed characters lead readers on an adventure as they discover that bubbles may grow and flow, fly in the sky and pop, but dipping the stick and blowing can make more.

Websites

Exploratorium Bubbles Page
www.exploratorium.edu/ronh/bubbles/bubbles.html

Bubblesphere
http://bubbles.org

Name: _____

Bubble Shapes
Data Table

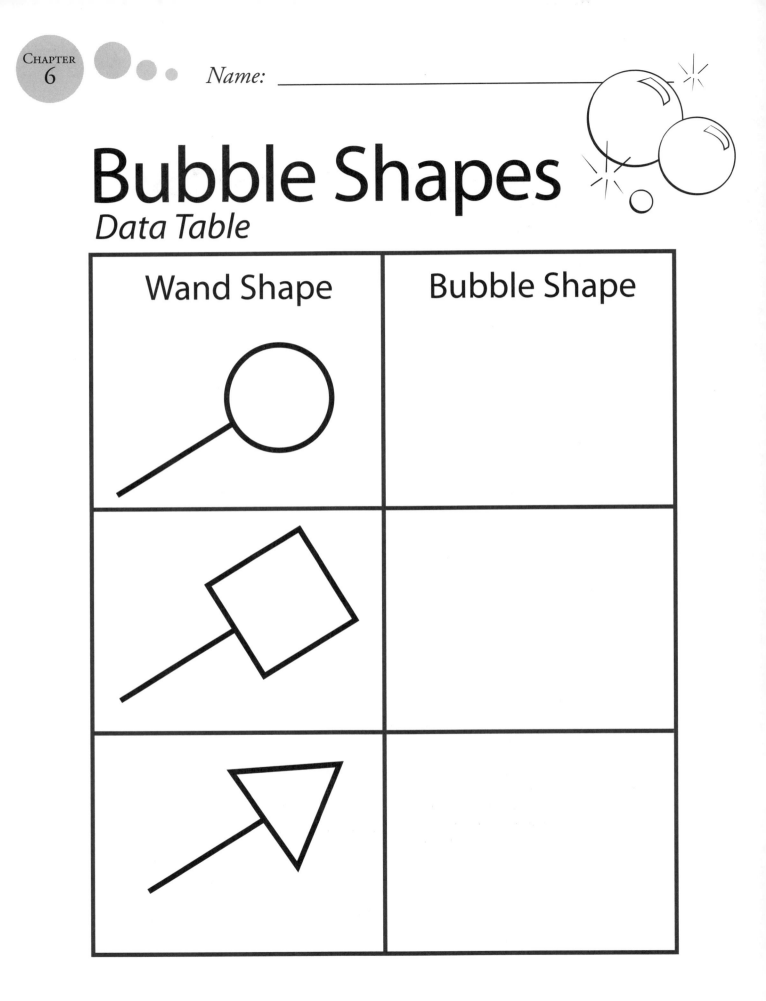

Wand Shape	Bubble Shape

How Big Is a Foot?

Description

Learners explore the history of measurement from the ancient Egyptian use of nonstandard units to the modern-day metric system. They learn why standard measuring tools are useful and that their development was a problem-solving process that took centuries.

Suggested Grade Levels: 2–4

Lesson Objectives Connecting to the Standards

Content Standard E: Science and Technology

- Understand that people have always had problems and invented tools and techniques (ways of doing something) to solve problems.
- Understand that tools help scientists make better observations and measurements for investigations. They help scientists see, measure, and do things that they could not otherwise see, measure, and do.

Content Standard G: History and Nature of Science

- Understand that science and technology have been practiced by people for a long time, and that men and women have made a variety of contributions throughout the history of science and technology.

Featured Picture Books

Title	*How Big Is a Foot?*	*How Tall, How Short, How Faraway*
Author	Rolf Myller	David A. Adler
Illustrator	Rolf Myller	Nancy Tobin
Publisher	Young Yearling	Holiday House
Year	1962	1999
Genre	Story	Non-narrative Information
Summary	The King has a problem. He wants to give the Queen a bed for her birthday, but no one knows the answer to the question "How big is a bed?"	Colorful cartoons and easy-to-follow text introduce the history of measurement, from the ancient Egyptian system to the metric system.

Time Needed

This lesson will take several class periods. Suggested scheduling is as follows:

Day 1: **Engage** with *How Big Is a Foot?* read aloud, **Explore** with Measuring With Feet,

and **Explain** with A Letter to the King.

Day 2: **Elaborate** with *How Tall, How Short, How Faraway* read aloud and Measurement Activities.

Day 3: **Evaluate** with A Better Way to Measure.

Materials

About 2 m of string or yarn per pair of students

Roll of masking tape

Yardstick

Meterstick

Rulers (1 per student)

Student Pages

A Letter to the King

A Better Way to Measure

Background

The National Science Education Standards suggest that children develop some essential understandings about science and technology, including the idea that people throughout history have invented tools and techniques to solve their problems. In this lesson, students learn how the development of standard measurements was a fascinating but lengthy problem-solving process.

Weights and measures were among the first tools invented by man. Ancient people used their body parts and items in their surroundings as their first measuring tools. Early Egyptian and Babylonian records indicate that length was first measured with the forearm, hand, and fingers. As societies evolved, measurements became more complex. It became more and more important to be able to measure accurately time after time and to be able to reproduce the measurements in different places. By the 18th century, England had achieved a greater degree of standardization in measurement than other European countries. The English, or *customary system* of measurement, commonly used in the United States, is nearly the same as that brought by the colonists from England.

The need for a single, worldwide measurement system was recognized more than 300 years ago when a French priest named Gabriel Mouton proposed a comprehensive decimal measurement system. A century passed, however, and no action was taken. During the French revolution, the National Assembly of France requested that the French Academy of Sciences "deduce an invariable standard for all the measures and all the weights." A system was proposed that was both simple and scientific: the *metric system*. The simplicity of the metric system is due to its being based on units of 10. The standardized structure and decimal features of the metric system made it well suited for scientific and engineering work, so it is not surprising that wide acceptance of the metric system coincided with an age of rapid technological development. By an Act of Congress in 1866, it became "lawful throughout the United

States of America to employ the weights and measures of the metric system to all contracts, dealings, and court proceedings." By 1900, a total of 35 nations had accepted the metric system. Eventually, the name *Systeme Internationale d'Unites* (International System of Units) with the international abbreviation *SI* was given to the metric system. Although the customary system of measurement is commonly used in everyday situations in the United States, American scientists primarily use the metric system (SI) in their daily work.

Adapted from: A Brief History of Measurement Systems
    http://standards.nasa.gov/history_metric.pdf#search='history%20of%20measurement'

engage

How Big Is a Foot? Read Aloud

Inferring

Show students the cover of the book *How Big Is a Foot?* Ask

? What can you infer from the title and illustration on the cover of this book?

Begin reading the book aloud, but stop after reading, "Why was the bed too small for the Queen?"

explore

Measuring with Feet

Remind the students that "the King took off his shoes and with his big feet walked carefully around the Queen. He counted that the bed must be three feet wide." Tell students that they are going to determine the length of three feet by using their own feet. Give each pair of students about 2 m of yarn or string. Then demonstrate the steps for measuring three "feet":

Measuring three "feet"

1 Have your partner hold the end of the string where the back of your heel touches the floor.

2 Place one foot right in front of the other for three steps, and then freeze.

3 Have your partner stretch the string to the big toe of your third step.

4 Cut the string. It will now represent the length of your three "feet."

5 Attach a piece of masking tape with your name on it to one end of the string.

6 Hang your string from the board.

7 Help your partner measure his or her three "feet."

When all students have hung their strings on the board, compare the various lengths. Ask

? Are all of the strings the same length? Why or why not?

Hold up a yardstick and ask

? How many feet are in a yard? (Students may know there are three feet in a yard.)

? How does your string compare to three feet as measured by a yardstick?

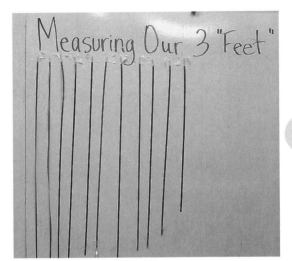

Comparing the strings

explain

A Letter to the King

Refer back to the book, *How Big Is a Foot?* Ask

? Do you think it is fair that the King put the apprentice in jail? Why or why not?

? What advice would you give the King about how he might be able to get a bed the right size for the Queen?

On A Letter to the King student page, have students write a persuasive letter to the King about why he should let the apprentice out of jail. Ask them to explain why the bed is too small for the Queen and what he could do to get a bed that is the right size. Have students share their letters with a partner. Read the student letters to assess

whether or not students understand that the bed is too short because of the difference in foot size between the King and the apprentice. Students should also provide reasonable advice for getting a bed the right size.

Next, read the rest of the book to students. Ask

? How does the apprentice's solution in the book compare to the advice you gave in your letter?

? Is there more than one correct solution? (yes)

? What are some other ways that the King could have had the bed built the right size?

elaborate

How Tall, How Short, How Faraway Read Aloud and Measurement Activities

Making Connections: Text to Text

Show students the cover of *How Tall, How Short, How Faraway.* Ask

? What do you think this book is about?

Hold up *How Tall, How Short, How Faraway* and *How Big is a Foot?* Ask

? What do you think these two books might have in common? (Examples of answers include that both are about measurement and about people's height.)

Ask students to signal when they hear or see any connections between the two books. Then read pages 1–19 aloud of *How Tall, How Short, How Faraway* (ending after "5,280 feet are 1 mile") stopping to discuss any text-to-text connections. Students may point out some of the following connections:

How Tall, How Short, How Faraway	How Big Is a Foot?
Ancient Egyptians measured with their hands and arms (digits, cubits, palms, spans).	The King measured with his feet.
Measuring with hands and arms caused problems with getting accurate measurements.	Measuring with different-sized feet caused problems in getting a bed the right size for the Queen.
In the past, people often used their leader's or king's cubit or steps as a standard.	The apprentice decided to use the King's foot as a standard to remake the bed.
People made measuring sticks the size of their king's cubit or steps.	The apprentice made an exact marble copy of the King's foot and measured with it.

Measurement Activities

Challenge students by asking

? Can you measure the length of my desk without a ruler?

As a class, brainstorm a list of ways that you could measure the desk without using any traditional measuring tools. Then ask

? How did the ancient Egyptians measure out a span? (A span is the distance from the tip of your thumb to the end of your little finger with your hand stretched wide.)

Make the following data table on the board:

Length of Desk		
Name	Spans	

Call on a student to measure your desk with his or her hand span. Write that student's name and his or her number of spans on the data table. Call on another student, who is either taller or shorter than the first student, to measure the desk in his or her spans. Write that number of spans on the data table. Then measure the desk using your own hand span, and write that number of spans on the data table. Ask

? Which is the correct answer for the length of my desk? (Students should begin to understand that there is no "correct" answer in spans.)

Measuring the desk in spans

? Why did we get different answers for the length of the desk in spans? (Each person's span is a different size.)

? Why do you think the span is no longer used for measuring length? (It is not an accurate measurement because the length of the span varies from person to person.)

Go back to *How Tall, How Short, How Faraway*, and read from page 20 ("The metric system was first proposed over 300 years ago …") to page 31 ("People have been measuring things for thousands of years."). After reading, state that in different times and parts of the world, there have been many systems of measurement. Ask

? What are the two systems of measurement most widely used today? (the customary, also known as the English system, and the metric system)

Explain that the units used in these systems are called standard units. Standard units are units of

measurement that are accepted and used by most people. Some examples of standard units are: feet, inches, pounds, centimeters, meters, grams, and liters. The other type of units is nonstandard units, which are everyday objects that can be used to obtain a measurement. Examples include: spans, cubits, paces, and digits.

Explain that most people around the world, as well as scientists everywhere, use the metric system because it is simpler and less confusing than the customary system. Explain that, although the metric system was invented over 200 years ago, the United States has not entirely switched over to it. Some metric units are: grams, kilograms, liters, centimeters, and meters. Ask

? What things do you know of that are measured in metric units? (Examples include 2 l bottles of soda, 100 m dash, grams of fat in food, distances in kilometers, and kilometers per hour on a speedometer.)

Finding something one meter long

Comparing spans and centimeters

Give each pair of students a meterstick and a metric ruler. Have students use these tools to find something in the room that is about a centimeter long and something that is about a meter long.

Next, label the third column on the data table "centimeters" and call on a student to measure your desk in centimeters with a meterstick. Write that student's name and the measurement on the data table. Call on another student to measure the desk with the meterstick. Write that measurement on the data table. Then measure the desk yourself with a meterstick and write that measurement on the data table.

? Why were the answers so different for the length of the desk in spans, yet all the same for the length of the desk in centimeters? (Each person's span is a different size, but a centimeter is always the same size.)

evaluate

A Better Way to Measure

Review what students have learned about the history of measurement and the need for standard measuring tools. Then distribute the assessment student page, A Better Way to Measure. Correct responses may include the following:

1 The pirates disagree because, when they measured the distance to the treasure, they each used their own paces. One is tall and one is short, so their paces are different lengths.

2 The pirates could measure in feet, yards, or meters.

3 We wouldn't know the exact distance to anything or anyplace.

Inquiry Place

Have students brainstorm testable or researchable questions such as

? Do taller students always have a longer pace?

? What is the difference between the longest pace and the shortest pace of students in the room?

? What is the relationship between hand span and height of students in the room?

? Which countries have not adopted the metric system as their official system of measurement? Why?

Have students select a question to investigate as a class, or have students vote on the question they want to investigate as a team. After they make their predictions, they can design an investigation to test their predictions. Students can present their findings at a poster session or gallery walk.

More Books to Read

Eboch, C. 2006. *Science measurements: How heavy? How long? How hot?* Minneapolis: Picture Window Books.

Summary: Big or small? Full or empty? Heavy or light? Hot or cold? Things are measured everyday. Learn about the many tools needed for different kinds of measurements in science and at home.

Jenkins, S. 2004. *Actual size.* New York: Houghton Mifflin.

Summary: With his colorful collage illustrations, Jenkins shows the actual sizes of many interesting animals. Some pages show the entire animal, while others show only a part of the animal.

Leedy, L. 1997. *Measuring penny.* New York: Henry Holt.

Summary: Lisa learns about the mathematics of measuring by measuring her dog Penny with all sorts of units, including pounds, inches, dog biscuits, and cotton swabs.

Nagda, A. W. and C. Bickel. 2000. *Tiger math: Learning to graph from a baby tiger.* New York: Henry Holt.

Summary: At the Denver Zoo, a Siberian tiger cub named T. J. is orphaned when he is only a few weeks old. The zoo staff raises him, feeding him by hand until he is able to eat on his own and return to the tiger exhibit. The story is accompanied by graphs that chart T.J.'s growth, showing a wonderful example of real-world mathematics.

Pluckrose, H. 1995. *Length.* London: Watts Books.

Summary: Photographs and simple text introduce the concept of length and ways to measure it.

Wells, R. 1993. *Is a blue whale the biggest thing there is?* Morton Grove, IL: Albert Whitman.

Summary: This fun, kid-friendly book describes the relative sizes of a blue whale, Mount Everest, the Earth, and ultimately the universe.

Wells, R. 1995. *What's smaller than a pygmy shrew?* Morton Grove, IL: Albert Whitman.

Summary: In this fun book about the smallest of the small, Wells compares the size of a tiny animal, a pygmy shrew, to an insect, which is in turn contrasted with one-celled animals, bacteria, molecules, atoms, and subatomic particles.

Name: _____

A Letter to the King

Write a letter to the King to convince him to let the apprentice out of jail. In your letter, be sure to
- explain why the bed was made too small for the Queen, and
- tell what the King could do to have a bed made the right size for the Queen.

Your Royal Highness,

_____

_____

_____

_____

_____

_____

_____

_____

_____

_____

_____

Your Loyal Subject,

_____

Name: _____

A Better Way to Measure
to Map a Buried Treasure!

1. Why do you think the pirates disagree about the distance to the treasure?

_____

_____

_____

2. What would be a better way to measure the distance to the treasure?

_____

_____

_____

3. What would happen if everyone used his or her own paces to measure distance or length?

_____

_____

_____

Hear Your Heart

Description

Learners explore how the heart works, how exercise affects heart rate, and how to keep the heart healthy through regular cardiovascular exercise.

Suggested Grade Levels: K–2

Lesson Objectives Connecting to the Standards

Content Standard A: Science as Inquiry
- Design and conduct simple experiments to answer questions.
- Use data to construct reasonable explanations.

Standard F: Science in Personal and Social Perspectives
- Understand that individuals have some responsibility for their own health, and that personal care, including exercise, will maintain and improve health.

Featured Picture Books

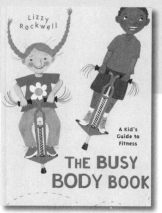

Title	*Hear Your Heart*	*The Busy Body Book: A Kid's Guide to Fitness*
Author	Paul Showers	Lizzy Rockwell
Illustrator	Holly Keller	Lizzy Rockwell
Publisher	HarperCollins	Crown
Year	2001	2004
Genre	Non-narrative Information	Non-narrative Information
Summary	This Let's-Read-and-Find-Out Science book provides a simple explanation of the structure of the heart and how it works. It includes diagrams and a few hands-on activities.	This action-packed introduction to the human body combines colorful illustrations of active kids with detailed information about how the body works.

Time Needed

This lesson will take several class periods. Suggested scheduling is as follows:

Day 1: **Engage** with Your Heart Anticipation Guide (before), and **Explore/Explain** with cardboard stethoscopes and *Hear Your Heart* read aloud.

Day 2: **Elaborate** with *The Busy Body Book* read aloud and graph.

Day 3: **Evaluate** with Your Heart Anticipation Guide (after).

Materials

Paper towel or toilet paper tubes (1 per pair of students)

Your Heart Anticipation Guide overhead

Student Pages

MVP (most valuable point)

Busy Body Cutouts (1 per student)

Background

It's a fact: Kids today are less fit than they were only a generation ago. Many are showing early signs of cardiovascular risk factors such as physical inactivity, excess weight, and higher blood cholesterol. Now more than ever, it is important to teach children healthy habits. The National Science Education Standards suggest that young children begin to realize that individuals should take some responsibility for their own health and that personal care (dental hygiene, cleanliness, and exercise) can help them maintain or improve their health. This lesson focuses on the important role of exercise in living a healthy life and, in particular, the way in which exercise benefits the heart.

The *heart* is a fist-sized muscle located between the *lungs* and slightly left of center in the rib cage. It contracts and relaxes automatically as it pumps blood to all parts of the body through an intricate system of *blood vessels*. A healthy heart makes a "lub-dub" sound with each beat. This sound comes from the *valves* shutting inside the heart. Blood leaves the left side of your heart and travels through blood vessels called *arteries,* which gradually divide into capillaries. Inside *capillaries* in the lungs, an oxygen-carbon dioxide exchange occurs and in the intestines, a nutrient-waste exchange occurs. The blood then travels in *veins* back to the right side of your heart, and the whole process begins again. The heart is so good at its job that it takes less than 60 seconds to pump blood to every cell in your body.

You can feel each time the heart squeezes a jet of blood into the arteries by finding your *pulse.* Two good places to find it are on the side of your neck just below the chin (the carotid artery pulse) and on the inside of your wrist just below the thumb (the radial artery pulse). When you are resting, you will probably feel between 70 and 100 beats per minute. A child's resting heart rate is faster than an adult's. Because the heart is a muscle, exercising it helps keep it healthy and strong. The American Heart Association recommends you do some sort of cardiovascular exercise for 30 to 60 minutes most days of the week. *Cardiovascular,* or aerobic, exercise is moderate exercise done for a long period of time that gets your heart rate up, such as running. Not smoking, eating a variety of healthy foods, and avoiding saturated and trans fats also help keep your heart healthy.

engage

Your Heart Anticipation Guide (Before)

Anticipation Guide

Begin by asking students

? What is the most important part of your body? Why do you think so?

Responses will vary, but tell students that they will be learning about a very important organ they couldn't live without: the heart. Using an overhead transparency of the Your Heart anticipation guide as a preassessment tool, read the statements about the heart and have the class discuss whether they think each statement is true or false. Mark answers in the "Before" column of the anticipation guide. At the end of the lesson, you will use student responses on the "After" column as an assessment.

explore/explain

Cardboard Stethoscopes and *Hear Your Heart* Read Aloud

Have students look at their closed fists. Tell them that the human heart is about the size of a closed fist. Have them place their fists against their chests and explain that the heart is located a little to the left of the middle of their chests. Have students quickly open and close their fists over and over again until their hands get tired. Explain that their fist-sized heart squeezes this way every second of every day without getting tired.

Ask

? What does your heart sound like? (One answer is a beating sound.)

? How does a doctor listen to your heart? (through an instrument called a *stethoscope*)

Give each pair of students a cardboard paper-towel or toilet-paper tube "stethoscope." They will be listening to each other's hearts through the stethoscopes. This can be a sensitive activity for

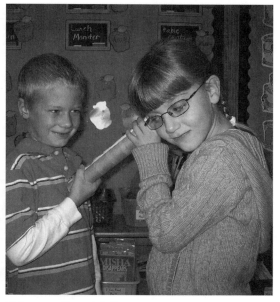
Using a cardboard stethoscope

some students. If a child seems anxious, suggest he or she listen to your heartbeat. Make sure the students are silent during this activity. Have them take turns listening to each other's heartbeats by putting one end of the tube on the left side of their partner's chest and placing their ear to the other end. After all students have had a chance to listen to their partner's heart, ask

? What does your partner's heart sound like? (It beats and it makes a "lub-dub" sound are two possible answers.)

? What activities might make your partner's heart sound differently? (Answers could include sleeping, running, and being afraid.)

? How do you think your partner's heart will sound after exercising?

Say, "Let's try it!" and have students stand to do some sort of cardiovascular exercise for one minute, such as marching or running in place while moving their arms up and down. You may want to play fast music during this part. Then have them repeat the cardboard stethoscope activity. Ask

? What does your partner's heart sound like after exercising? (It beats louder or faster or stronger.)

Marching in place to increase heart rate

? How do you think exercising your body helps your heart? (Answers will vary. They might include that the heart is a muscle and it needs to be exercised or that it beats faster when you exercise your body which also exercises your heart.)

Next, show students the book *Hear Your Heart* and explain that it is a nonfiction book that can help them find out more about the heart. Read the book aloud, skipping pages 12–21 to focus on the portions of the book specifically about the heartbeat and the pulse.

Synthesizing

After reading, ask

? Is the heart a muscle? (yes)

? How do you keep your muscles strong? (Exercise them.)

? What do you think you can do to keep your heart strong? (Exercise it by running, jumping, or playing sports are possible answers.)

? How often do you think you should do these

activities every week? (Answers will vary, but explain that they can keep their hearts strong by keeping their bodies busy by engaging in fast-paced physical activity for 30–60 minutes most days of the week.)

Explain that exercising regularly is important in keeping your heart healthy, as are eating healthy foods and not smoking.

elaborate

The Busy Body Book
Read Aloud and Graph

Determining Importance: MVP (Most Valuable Point)

Show students the cover of *The Busy Body Book* by Lizzy Rockwell, and tell them that, as you read, you would like them to think about the most important, or valuable, point they hear. Then read the book aloud. You may want to point out the labeled diagrams of the skeleton, the muscles, and other parts of the body without reading the entire diagram.

After reading, ask students to determine the *most valuable* point they learned from the book. Pass out the MVP (most valuable point) graphic organizer. Have students write what they feel is the most valuable point in the space provided. They can write VIPs (*very important points*) on the arms and legs. Have students discuss their writing with partners or small groups.

Next, make a large "busy body" graph by labeling the x-axis "Favorite Physical Activities" and the y-axis "Number of Students." Brainstorm a list of five or six cardiovascular activities that students enjoy and write them along the x-axis. Pass out "busy body" cutouts, and have students write their name and their favorite activity from the choices listed.

Then have students tape their cutouts on the graph to create a pictograph of favorite activities and analyze the graph together by asking questions such as

? What observations can you make about the graph?

? Which activity do students enjoy the most? The least?

? How many more people chose _____ than _____?

Remind students that they can keep their hearts strong by keeping their bodies busy with fast-paced physical activity for 30–60 minutes most days of the week.

evaluate

Your Heart Anticipation Guide (After)

Revisit the Your Heart anticipation guide overhead. Have the class vote again on whether each statement is true or false, and mark the answers

Favorite physical activity

Making a busy-body graph

in the "After" column. Discuss evidence for each statement from the activities and the reading, such as, "Our hearts are muscles so they need exercise like our other muscles."

The answers to the Anticipation Guide are

1 False

2 True

3 True

4 False

5 True

6 True

You can assess students individually by having them write about what their heart does for their body and what they can do to keep it healthy. Have them include a drawing of themselves taking part in their favorite cardiovascular activity.

Inquiry Place

Have students brainstorm testable questions about heart rates such as

? What is the average heart rate of the class?

? Are boys' heart rates faster than girls'?

? In which activity is your heart beating the fastest: playing a handheld video game, jogging, or jumping rope?

Have the class vote on its favorite question, and then discuss how to investigate the question. Teach students how to measure their heart rates. To do this, have them sit very still with elbows slightly bent and palms upturned, and then gently place the middle and index fingers of one hand on the other wrist at the base of the thumb. They may have to try a few different spots until they feel a soft beating. Tell them this beating is called their *pulse* and it is caused by the heart squeezing blood through their body. (If students are unable to locate the pulse at their wrists, they can try gently placing their middle and index fingers against one side of the neck, just below the chin.) Tell students that their pulse can tell them how fast their heart is beating if they count the number of beats for one minute. To save time, have students practice taking their pulses during six-second intervals. Using a clock with a second hand to keep track of time, say "start" to initiate the count and "stop" after six seconds have passed. Tell students they can multiply by 10 (or just add a zero to their count) to figure out how many times their hearts beat in one minute.

Next, introduce the idea of controlling an experiment by asking the students to think about how to make it a "fair" test. Guide the students in designing a procedure and collecting and analyzing the data. Have teams of students formulate conclusions (answer the question based on the data) and present their findings at a poster session or gallery walk.

More Books to Read

Cole, J. 1990. *The magic school bus inside the human body.* New York: Scholastic Press.

Summary: Mrs. Frizzle and the gang board the bus for a wild ride through the human body. The familiar dual-purpose format, with informative insets accompanying the story, entertains while educating readers about digestion, the blood and heart, the brain, and more.

Seuling, B. 2002. *From head to toe: The amazing human body and how it works.* New York: Holiday House.

Summary: Inviting, straightforward text accompanied by cartoonish illustrations takes readers on a journey through the human body. Includes descriptions of the skeleton, muscles, brain, skin, and other organs.

Showers, P. 2004. *A drop of blood.* New York: Harper-Collins.

Summary: This engaging introduction to the composition and functions of blood features amusing illustrations of a Dracula-like vampire and his Igorish friend. Includes diagrams and sharp scanning electron micrographs of blood cells.

Simon, S. 1996. *The heart: Our circulatory system.* New York: Mulberry Books.

Summary: This fact-filled book describes in detail the heart, blood, and other parts of the circulatory system and explains in detail how each component functions. Vivid computer-enhanced photographs and micrographs, taken by scanning electron microscopes, accompany the text.

Sweeny, J. 2000. *Me and my amazing body.* Dragonfly Books.

Summary: In this simple picture book for younger children, the narrator takes readers on a guided tour through her "amazing body." Accompanied by colorful cartoon art, the book explains human anatomy and physiology in an engaging and accessible way.

Website

American Heart Association
www.americanheart.org

Your Heart

Anticipation Guide

Before
True or False

After
True or False

_____ 1. Your heart is closest to the right side
of your chest. _____

_____ 2. Your heart is about as big as your fist. _____

_____ 3. Your heart is a strong muscle. _____

_____ 4. Your heart beats slower when
you exercise. _____

_____ 5. A big animal's heart beats more
slowly than a little animal's heart. _____

_____ 6. You should exercise for 30 to 60
minutes most days of the week. _____

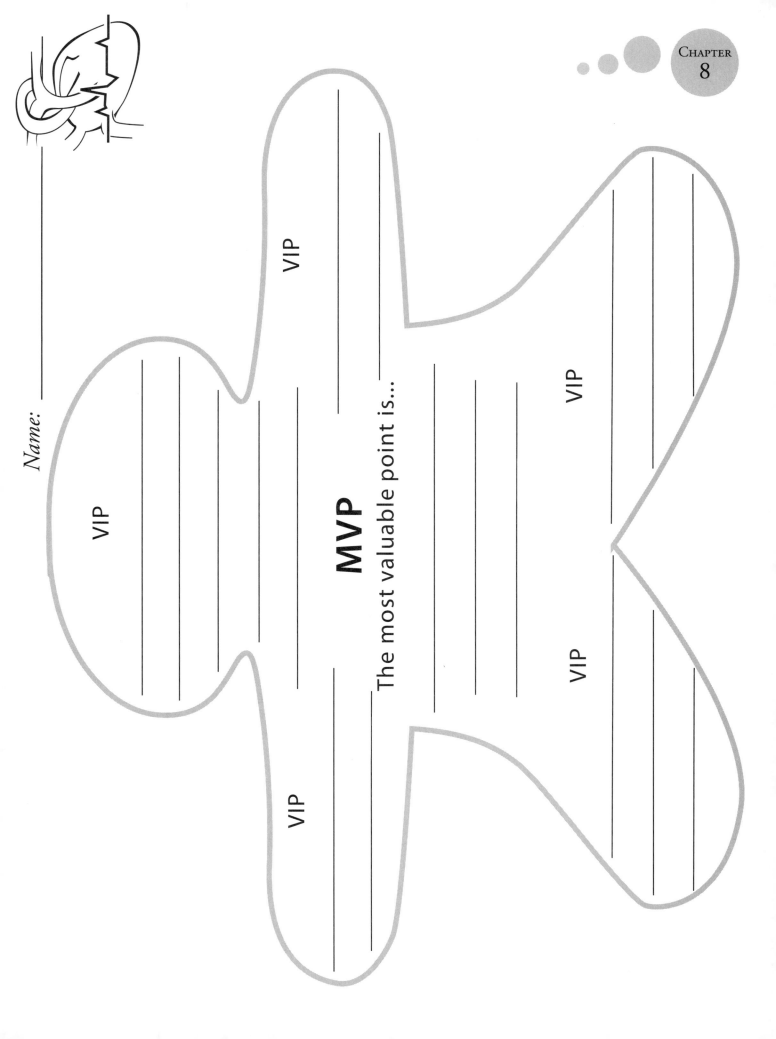

Name: _____

MVP
The most valuable point is...

VIP

VIP

VIP

VIP

VIP

VIP

Busy Body Cutout

Busy Body Cutout

Loco Beans

Description

Learners explore mystery objects and discover that they are Mexican jumping beans: seed pods containing moth larvae. They learn about the life cycle of the jumping bean moth and conduct a simple experiment to see how temperature affects the rate at which the larvae "jump."

Suggested Grade Levels: 3–4

Lesson Objectives Connecting to the Standards

Content Standard A: Science as Inquiry

- Ask a question about objects, organisms, and events in the environment.
- Design and conduct simple experiments to answer questions.
- Use data to construct reasonable explanations.

Content Standard C: Life Science

- Understand that animals have basic needs: air, water, and food. Animals can survive only in environments in which their needs can be met, and distinct environments support the life of different types of animals.
- Understand that animals have life cycles that include being born, developing into adults, reproducing, and eventually dying.

Featured Picture Book

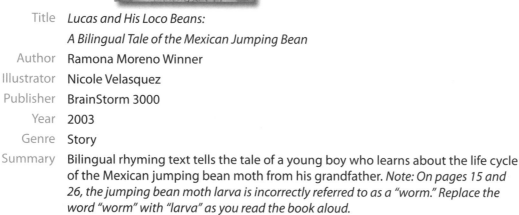

Title	*Lucas and His Loco Beans: A Bilingual Tale of the Mexican Jumping Bean*
Author	Ramona Moreno Winner
Illustrator	Nicole Velasquez
Publisher	BrainStorm 3000
Year	2003
Genre	Story
Summary	Bilingual rhyming text tells the tale of a young boy who learns about the life cycle of the Mexican jumping bean moth from his grandfather. *Note: On pages 15 and 26, the jumping bean moth larva is incorrectly referred to as a "worm." Replace the word "worm" with "larva" as you read the book aloud.*

Time Needed

This lesson will take several class periods. Suggested scheduling is as follows:

Day 1: **Engage/Explore** with Mystery Object O-W-L, and **Explain** with *Lucas and His Loco Beans* read aloud.

Day 2: **Elaborate** with Loco-Motion Investigation and Finding the Median.

Day 3: **Explain** with The Mexican Jumping Bean article.

Day 4: **Evaluate** with Loco Beans Poster.

Materials

Mexican Jumping Beans (1 per student)

> In advance, open the jumping bean package and place seed pods on a flat surface beneath a lamp. As they warm up, they will quickly become active and you will be able to determine which seed pods contain live moth larvae. Collect the most-active seed pods to use for this lesson, as some students may become discouraged if their "beans" don't jump!

Petri dishes or clear plastic cups (1 per student)

Hand lenses (1 per student)

Centimeter rulers (1 per student)

Think pads (sticky notes)

Heating pads or desk lamps (1 per team of 4–6 students)

Alcohol thermometers (1 per team of 4–6 students)

Clock with a second hand

Loco-Motion Class Data overhead

SAFETY

Warning:
Jumping beans are a choking hazard for children under four.

Jumping beans are available from

Jumping Beanditos
www.jbean.com

My Pet Beans
www.mypetbeans.com

JumpingBeansRUS.com
www.jumpingbeansrus.com

Student Pages

O-W-L student page

Loco-Motion and Loco-Motion Class Data

The Mexican Jumping Bean article

Loco Beans Poster Rubric

Background

This lesson addresses both the Life Science Standard and the Science as Inquiry Standard from the National Science Education Standards. The Standards suggest that as students in grades K–4 develop scientific concepts and vocabulary, they should also develop inquiry skills. This lesson is designed not only to teach students about the life cycle of the Mexican jumping bean moth but also to teach students to focus on the processes of doing scientific investigations.

Mexican jumping beans are not really beans. They are actually sections of the *seed pods* (Figure A) of a shrub native to the mountains of Mexico. In early summer, small moths emerge from the seed pods, mate, and deposit their eggs on the flowers of the host plant. When an egg hatches, the tiny caterpillar *larva* burrows its way into the seed pod. Once inside, it lines its new home with silk and begins eating the developing seeds. The larva receives water that is absorbed when it rains or during heavy dew. Air is also taken in through the seed pod.

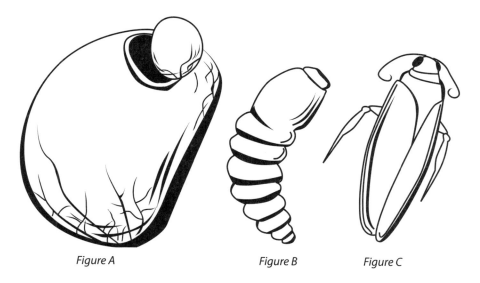

Figure A   Figure B   Figure C

Later in the summer, the seed pods fall to the ground and split into three separate sections. The sections containing moth larvae begin jumping and rolling about in response to temperature and possibly other factors. It is thought that they move to escape the heat, perhaps positioning themselves into a cooler spot where they can safely pupate. To move, a larva grasps the silken lining of the seed pod with its forelegs and jerks its body. As the seed pods move around on the dry ground, they sound like pattering rain drops! If the seed pods are shaken, they will temporarily stop moving.

The seed pods remain still during the winter months as the larvae transform into *pupae*. The following summer, each pupa (Figure B) pushes through a small circular door it has previously made in the wall of the seed pod, the pupal covering splits open, and a small gray moth (Figure C) emerges. It lives only a few days to mate and lay eggs. The cycle then begins again.

The U.S. Department of Agriculture permits importation of jumping beans into the United States, because the moths cannot infest local plant species. These activities are best done in late summer or fall when active seed pods are available for ordering. If kept at a cool temperature (above freezing), and lightly misted with water each week, jumping bean seed pods can be easily maintained in the classroom for several months. Some vendors advise soaking them in dechlorinated water for 5 hours once a month instead of misting. Drying your beans is very important. Leaving them wet may cause mold to grow on them, shortening their life expectancy. Most of the larvae will die in the seed pods, but, if they do hatch, the moths are harmless and will live only a few days.

engage/ explore

🐛 Mystery Object O-W-L

In this activity, students will be given a mystery object to observe: a jumping bean seed pod. Some students may already be familiar with Mexican jumping beans and want to tell others what they are. To help prevent students from "spilling the beans," you can say, "I am going to give you a fascinating mystery object. Some of you may have seen these before. If you have and you know what they are, please don't share with the class just yet. Give your classmates the chance to unravel the mystery on their own."

Then give each student a mystery object (jumping bean seed pod) in a petri dish or clear cup. Ask them to spend a minute or two observing the mystery objects using all of their senses except

Observing the mystery objects

taste. Explain that, when scientists observe things, they record their observations by both drawing and writing. Give each student a copy of the O-W-L (Observe/Wonder/Learn) student page, a hand lens, and a metric ruler. Have students draw the mystery objects and then record their observations, including measurements. When they begin noticing the movement of the seed pods, students may become so engaged that they stop writing down observations. You may want to say, "See how many observations you can write in one minute!" and time them.

Next, have students list their wonderings about the objects. Ask students to share some of their wonderings out loud. Then allow them to share inferences about the identity of the objects with a partner.

Collect the seed pods for use on Day 2 for the Loco-Motion investigation. After the investigation, students will enjoy taking a jumping bean home with them.

explain

Lucas and His Loco Beans Read Aloud

In advance, use blank paper to hide the cover of the book *Lucas and His Loco Beans* and don't tell students the title of the book. Use a small piece of sticky note paper to replace the word *worm* with the word *larva* on page 15 so that students don't get misconceptions from the book. (The jumping beans contain insect larvae, which are in a different classification than worms.) Make a T-chart with the selected Spanish vocabulary words from the book on the left side. Leave the right side blank—you will fill it in as you read. Make sure all students have think pads (sticky notes) to write on. They will be using them to jot down their inferences about the meanings of some of these Spanish words:

Spanish	English
abuelo	
verano	
lugar	
semillas	
polillas	

Determining Importance

Say, "I have a very interesting storybook to share that may help you figure out what the mystery objects are. I have hidden the cover of the book so that you will need to listen very carefully for clues as I read. You will hear some words you might not know, such as *abuelo, verano, lugar, semillas,* and *polillas.* You will also use clues from the reading to figure out the meanings of these words." Then begin reading aloud *Lucas and His Loco Beans* through page 24 using the inferring strategy described on the next page. Remember to replace the word *worm* with *larva* when you read page 15.

Inferring

Read page 3 (*"Abuelo, Abuelo,* I yelled as I ran, I hugged him and held him, grabbed hold of his hand."*) and show the picture, then ask if students can figure out what the Spanish word *abuelo* means. Have them write their inferences on their think pads.

Ask

? What do you think *abuelo* means? (Responses will vary.)

Explain that when we read an unfamiliar word, we can infer, or figure out, its meaning by using clues from the text and illustrations. We can also use our prior knowledge, or schema, to infer the meanings of words.

Ask

? How did you infer the meaning of *abuelo*? (From the story, from the picture, and from past experiences are some of the likely answers.)

Tell students that *abuelo* means grandfather in Spanish, and then write the translation on the T-chart. Continue reading, stopping to have students jot down their inferences about the meanings of the selected Spanish words. After brief discussion about how they inferred the meanings, write the English translations on the chart. (You may want to translate the Spanish words not listed on the chart as you read, without stopping to have students write inferences.) Continue reading the book up to page 25. The information on pages 26–29 will be used later in this lesson.

After reading page 12, students will be able to infer that the mystery objects are *son brincadores* or Mexican jumping beans. They will also learn that the beans contain larvae that eventually develop into moths and that warmth from your hand makes the larvae move like crazy. After reading, have students write their learnings about the mystery objects in the "L" column of the O-W-L chart. Then have them explain their understandings about Mexican jumping beans with a partner. They can also record any more wonderings they

might have about the mystery objects after the read aloud.

Inferring the meaning of words in Spanish

Questioning

Then ask

? What do you think a good title would be for the book? (Answers will vary; show cover and title after students share ideas.)

? What are the mystery objects? (seeds or seed pods)

? Where do they come from? (from the Mexican jumping bean plant found in Mexico)

? What is inside them? (moth larvae)

? What do you think *loco* means? (crazy)

? Why do you think Lucas' grandfather calls them "loco beans"? (They act crazy by jumping around.)

elaborate

Loco-Motion Investigation and Finding the Median

Ask students

? What kinds of things might make the moth larvae move around more? (Brainstorm a list of variables that might affect the movement of the seed pods, such as light, heat, sound, and vibration.)

Next, reread page 19, "Your warm body makes them move like crazy," and ask

? How could you test the effect of heat on jumping bean movement?

Then tell students that you have designed an experiment to answer the question, "Do jumping bean larvae jump more at room temperature or at a warmer temperature?" Pass out the Loco-Motion student page, and have students work in groups to complete the investigation. You will need several desk lamps or heating pads placed at stations around the room so that students can warm their seed pods. Make sure they understand that excessive heat will kill the larvae. Do not heat them for more than a few minutes.

Procedure:

1 Have students use alcohol thermometers to measure the room temperature and the temperature beneath a desk lamp or on a heating pad. They can record both temperatures on their Loco-Motion student page.

2 Have students circle their predictions.

3 Discuss what will constitute a "jump." You may want to allow any movement (a wiggle, flip over, or jump) to count as a jump. Explain that everyone should record their data the same way so that you can compare each other's results fairly. Designate a student from each team to act as a timekeeper. (You may prefer to do this as a whole-group activity and act as the timekeeper yourself.) When the clock starts, students can begin collecting data by making a small tally mark on the back of their papers each time they detect a movement. When the minute is up, they can count up their tally marks and record that number in the data table. Next, have students place their seed pods beneath a lamp or on a heating pad for about one to two minutes. Then repeat the procedure.

4 Conclusion: Have students look at their own data and answer the question, "Did your jumping bean larva jump more at room temperature or a warmer temperature?"

5 Explain that an important part of doing a good investigation is comparing your data to other scientists' findings. Discuss whether everyone in the class had the same conclusion. If not, discuss reasons why, such as some seed pods never moved at all and some people recorded wiggles as movements while others only recorded jumps.

Finding the Median: Background for Teachers

The Loco-Motion investigation may result in the data set containing some extreme values, or *outliers*. For example, some students may report zero jumps while others may report more than a hundred jumps. In this case, it will be useful to report the class data as a *median* rather than a *mean*. The median value is a measure of central tendency; i.e., the middle value in a set of values. It is a summary statistic that provides us with a description of the entire data set and is especially useful with large data sets where we might not have the time to examine every single value. The median is less sensitive to extreme values than the mean, so it is a very good summary statistic to use when the data contains outliers. It is fairly simple for elementary students to calculate the median in a set of numbers using the procedure described.

Finding the Median:
Student Procedure

1 Tell students that it is often useful to look at the data of the entire class together. One way to do this is the find the median, or middle number, of the data. Have each student write his or her total number of jumps at room temperature on a sticky note.

2 Then have students come up to the board a few at a time to place their numbers in a row across the board from smallest to largest.

3 Look at the row of numbers on the board. Ask students to predict what they think the median, or middle number, will be. Explain that half of all the numbers on the board will be smaller than the median and half will be larger.

4 Choose two students to come to the board and stand at each end of the list. At the same time, have them remove the sticky note at each end of the list and continue removing the end values until just one or two sticky notes remain.

5 If only one sticky note remains, that is the median value. When the data set contains an odd set of numbers, the middle value is the median value.

6 If two sticky notes remain, then the median is the number in the middle of the two values.

7 Next, find the median for the "Warmer Temperature" data set using the sticky note method.

8 As a class, determine the range of the y-axis scale (0 to what number?). Graph both median values on the Loco-Motion Class Data overhead and discuss.

Using their data, students should find that, in general, Mexican jumping bean larvae move more at warmer temperatures. Then tell students that you have some other wonderings about jumping beans, such as

? Is a jumping bean really a bean?

? How do the larvae get inside the seeds?

Then ask students

? What questions do you have about jumping beans?

Students should list further questions in the "W" column of the O-W-L student page and share them with the class.

Finding the median

explain

The Mexican Jumping Bean Article

At this point, students will still have some unanswered questions about the jumping beans. Pages 26–29 of *Lucas and His Loco Beans* contain more information about the jumping bean moth life cycle, as well as diagrams. We have summarized this information in an article called "The Mexican Jumping Bean," so that students may read it in pairs and on their own.

Determining Importance/ Pairs Read

Tell students that you have an interesting article they can read to find out why the jumping bean larvae move more at warmer temperatures. They can also find out the answers to many of their questions about "loco beans." Pass out the article and have students do a *pairs read* by taking turns reading aloud the paragraphs. In a pairs read, one student reads aloud while the other listens and then summarizes the main idea. Benefits include increases in reader involvement, attention, and collaboration and making students become more independent learners. When students are finished, they can go back to their O-W-L pages and add their learnings to the "L" column.

evaluate

Loco Beans Poster

Give each student a copy of the Loco Bean Poster rubric, and have students or teams of students create a poster summarizing what they have learned about the Mexican jumping bean moth. Posters should include:

4 Points: Four facts about loco beans

3 Points: A detailed, labeled diagram showing the life cycle of the Mexican jumping bean moth

2 Points: A description of the Loco-Motion investigation and the team's conclusion

1 Points: An advertisement explaining what loco beans are and why they would make good pets

Extra Credit: A poem, song, rap, jingle, or "training tips" for the pet beans

Students can share their posters in a poster session or gallery walk. You can use the rubric to score completed posters and make comments.

Inquiry Place

Have students brainstorm testable questions such as

? What effect does sound have on jumping bean larvae?

? What effect does light have on jumping bean larvae?

? What effect does vibration have on jumping bean larvae?

Have students select a question to investigate as a class, or have groups of students vote on the question they want to investigate as teams. After they make their predictions, they can design an experiment to test their predictions. Students can present their findings at a poster session or gallery walk.

More Books to Read

Kalman, B. 2005. *Insect life cycles*. New York: Crabtree.

Summary: This nonfiction book explores the life cycles of a variety of insects in an engaging, fact-filled format. Includes full-color photographs, table of contents, bold-print words, captions, and a glossary.

Kalman, B. 2005. *Metamorphosis: Changing bodies*. New York: Crabtree.

Summary: This nonfiction book explains complete metamorphosis and incomplete metamorphosis by featuring a variety of organisms, including butterflies, ladybugs, frogs, dragonflies, and grasshoppers. Includes full-color photographs, table of contents, bold-print words, captions, and a glossary.

Himmelman, J. 1999. *A luna moth's life*. New York: Children's Press.

Summary: Simple text and vivid, close-up illustrations describe the daily activities and life cycle of the luna moth using a story format.

Name: _____

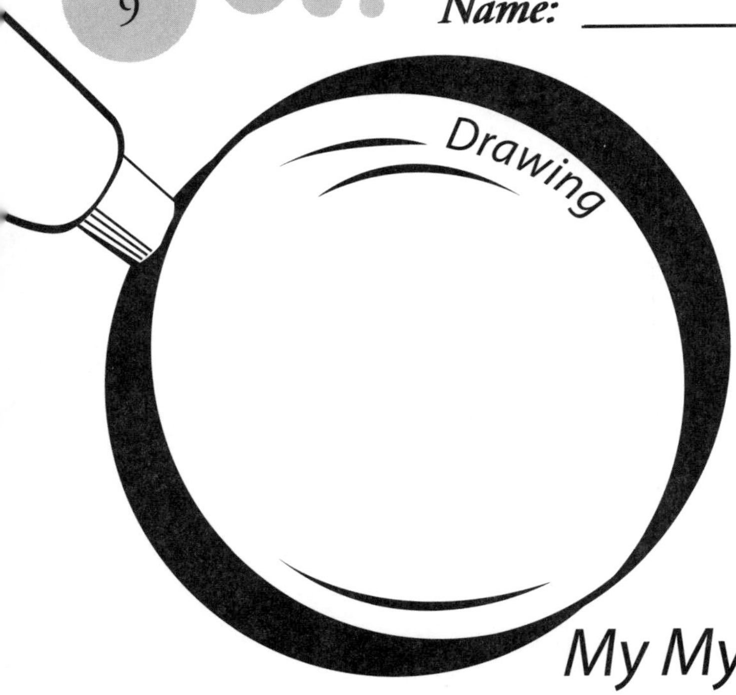

Drawing

My Mystery Object

O   W   L

What do you OBSERVE about the object?	What do you WONDER about the object?	What did you LEARN about the object?

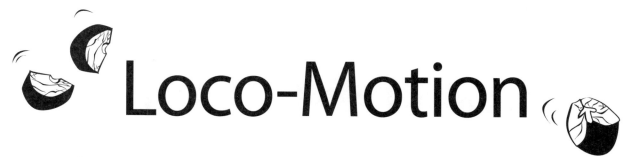

Loco-Motion

Do jumping bean larvae "jump" more at room temperature or at a warmer temperature? Let's experiment to find out!

1 Measure the temperature:

      Room temperature ___°C   Warmer temperature ___° C

2 Make a prediction (circle one): The jumping bean larva will jump more at

      Room temperature   -OR-   Warmer temperature

3 Collect data: Count the number of times your jumping bean larva jumps in one minute at room temperature and record on the data table below. Then count the number of times your jumping bean larva jumps in one minute at a warmer temperature and record on the data table below.

Data Table	
# of Jumps at Room Temperature	# of Jumps at Warmer Temperature

4 Conclusion: Look at the data table. Did your jumping bean larva jump more at room temperature or warmer temperature?

_____

5 Compare to other scientists' findings: Did everyone in the class have the same conclusion? Why or why not?

_____

_____

Name: _____

Loco-Motion
Class Data

Room Temperature Median # of Jumps	Warmer Temperature Median # of Jumps

"Loco-Motion" Class Bar Graph

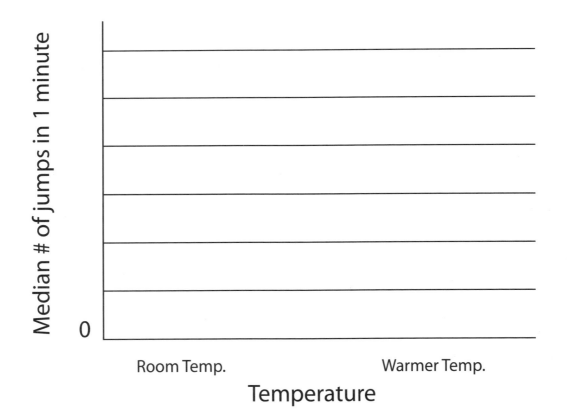

The Mexican Jumping Bean

What Is a Jumping Bean?

A Mexican jumping bean is not really a bean. It is part of the **seed pod** of a shrub found in Mexico. Inside the seed pod lives the **larva**, or caterpillar, of a small moth. The larva finds everything it needs to stay alive inside the seed pod: food, water, air, and shelter. It gets food by eating the inside of the seed pod. It gets water and air that enter the thin shell of the seed pod. The seed pod also gives the larva protection from weather and predators.

How Does It Get Inside?

In the early summer, a female moth lays her eggs on the flowers of a certain shrub found in the mountains. No other kind of plant will do. When the eggs hatch, the tiny white caterpillars chew their way into the seed pods of the shrub. The seed pods fall to the ground and split into three sections. Each section can contain a single larva. Not all of the sections contain larvae. If they did, there would be no seeds left to grow into Mexican jumping bean plants!

What Makes It Jump?

The larva spins silk to line the inside of its new home. When it gets too hot, it grabs onto the silk and snaps its body to make the seed pod jump and roll. When the larva reaches a cooler spot in the shade, it doesn't move as much. If you could hear the seed pods moving around on the dry ground, you might think they sound like rain drops.

How Does It Become a Moth?

The larva spends the summer and fall eating and moving around inside its seed pod. It chews a tiny flap in the seed pod called an exit hole. As winter approaches, it begins to spin more silk until it is covered with a soft **cocoon**. The larva begins to go through many changes. It is now called a pupa. The pupa slowly turns into an adult moth. These changes are known as **metamorphosis**. The following summer, the adult moth hatches out of the cocoon and squeezes through the exit hole it made months earlier. Soon it will find a mate, lay eggs, die, and the cycle will begin again.

Loco Beans
Poster Rubric

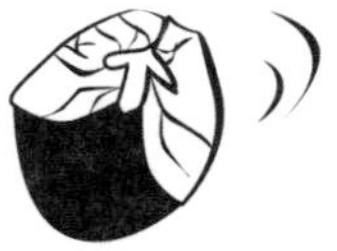

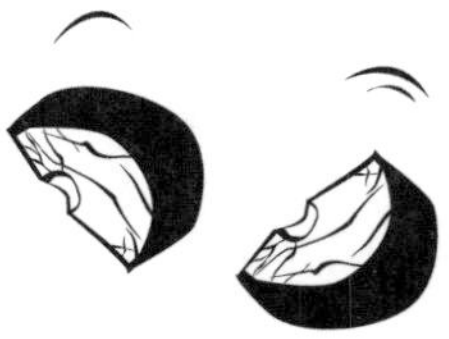

Your poster includes:

4 Points: Four facts you learned about loco beans

4 3 2 1 0

3 Points: A detailed, labeled diagram showing the life cycle of the Mexican jumping bean moth

3 2 1 0

2 Points: A description of the Loco-Motion investigation and your team's conclusion

2 1 0

1 Point: An advertisement explaining what loco beans are and why they would make good pets

1 0

Extra Credit: Your poster includes a poem, song, rap, jingle, or "training tips" for your pet beans.

1 0

Total Points_____/10

Comments: _____

Wiggling Worms

Description

Learners keep careful records in journals as they ask questions about earthworms, observe their adaptations, conduct simple experiments, and explore ways that earthworms help the Earth.

Suggested Grade Levels: 2–4

Lesson Objectives Connecting to the Standards

Content Standard A: Science as Inquiry

- Ask a question about objects, organisms, and events in the environment.
- Design and conduct simple experiments to answer questions.
- Use data to construct reasonable explanations.

Content Standard C: Life Sciences

- Understand that each plant or animal has different structures that serve different functions in growth, survival, and reproduction.
- Understand that organisms cause changes to their environments. Some of these changes are detrimental, and some are beneficial.

Featured
Picture
Books

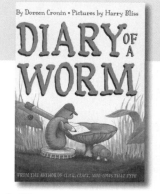

Title	*Diary of a Worm*	*Wiggling Worms at Work*
Author	Doreen Cronin	Wendy Pfeffer
Illustrator	Harry Bliss	Steve Jenkins
Publisher	HarperCollins	HarperCollins
Year	2003	2003
Genre	Story	Non-narrative Information
Summary	A young worm writes a hilarious journal about his daily adventures.	Explains how earthworms eat, move, and reproduce and how they help plants grow.

Time Needed

This lesson will take several class periods. Suggested scheduling is as follows:

Day 1: **Engage** with *Diary of a Worm* read aloud and Worm Wonderings.
Day 2: **Explore/Explain** with Earthworm Observations.
Day 3: **Explore/Explain** with Damp or Dry? Experiment and A Day in the Life of an Earthworm.
Day 4: **Explain** with *Wiggling Worms at Work* read aloud.
Day 5 and beyond: **Elaborate** with How Do Worms Help the Earth?
Day 6 and beyond: **Evaluate** with Save the Worms Posters and optional Wiggling Worms quiz.

Materials

Per class:
 Prepared worm bin with a ventilated lid and sides covered by black paper (see background for suggestions)

 Worms (night crawlers or red wigglers—see safety box)

Per pair for Earthworm Observations:
 Night crawler or other large earthworm
 Paper plate
 Paper towel
 Hand lens
 Metric ruler

Per group or class for Damp or Dry? Experiment:
 4 earthworms
 "Experimental chamber" made of a plastic shoebox with a lid and paper covering the sides, or a cardboard shoebox and lid

 Wet paper towel and dry paper towel

Per group or class for How Do Worms Help the Earth?
 2 flowerpots
 Soil from a yard
 Castings from outside a worm's tunnel
 Seeds (pumpkin, watermelon, grass seeds, lima beans, or other similar seed)

Per student or group for Save the Worms Posters:
 Poster board
 Markers

SAFETY

The worms should come from a commercial source that can guarantee they have been raised in a safe, nonchemical soil or medium. Students should wash their hands with soap and water after handling earthworms. Students who have cuts or hangnails or who are immune compromised should use gloves for this activity.

Also take care that the worms do not become dehydrated, are not too hot or cold, and are not handled roughly. Remind students that all living things should be handled gently.

Student Pages

My Worm Journal (Copy the cover on one sheet; copy pp. 2 and 3 back-to-back with 4 and 1. Fold and staple along spine.)
How Do Worms Help the Earth? (Make one-sided copies as students will be cutting out the cards.)
Save the Worms Poster Rubric
Wiggling Worms Quiz (optional)

Background

According to the National Science Education Standards, elementary students should build understanding of biological concepts through direct experience with living things. Earthworms are an ideal animal for this type of exploration. They are fascinating, easy to find, and fairly easy to care for. Earthworms are *annelids,* a phylum of animals that have joined, segmented body parts. There are thousands of different species of earthworms. Night crawlers are very large earthworms that eat soil and plant material. They are not the same as red worms, which are used for composting.

Earthworms literally eat their way through the earth. As they eat, they form *tunnels* that help aerate the soil and increase the rate of water movement into it. They also crawl above ground occasionally, pulling dead leaves and bits of plants back down with them. Earthworms leave behind droppings (known as *castings*) that make excellent fertilizer. A pile of worm castings located outside an earthworm's burrow is known as a *midden.* Soil is greatly enriched by the actions of earthworms.

Students can observe the external anatomy of the earthworm with the naked eye or a hand lens. Earthworms have many *adaptations* (body parts or behaviors that help an animal meet its needs) that enable them to lead a burrowing life. Students can observe and attempt to count the many *segments* that make up the long, tubelike body (night crawlers have about 150 segments). Each segment has eight small *bristles* that worms use to grip the ground as they move. If you place an earthworm on your arm, you can actually feel its bristles stick to your skin as it stretches and moves. Earthworms have both a *head* and a *tail.* The earthworm's head is closest to its thickened end, called the *saddle* (or *clitellum*). It usually crawls headfirst. The covering of its *mouth* serves as a wedge to force open cracks in the soil. A worm lacks teeth, so, when it swallows soil and organic material, that is passed to a storage area known as the *crop* and then ground up in the muscular *gizzard.* The food is then passed through the rest of its digestive system to the *anus,* where the castings are released back into the soil. Worms have no lungs. Gases are exchanged directly through the skin, which must be moist. That's why earthworms avoid dry ground, sun, and extreme heat. Worms lack eyes, ears, and noses, but do have taste and light receptors. They can also sense vibrations, which helps them hide when predators approach.

You can make a simple classroom worm bin using a clear plastic storage container, large glass jar, or small aquarium. It should be something that has a secure ventilated lid or cover to let air circulate but prevent the worms from climbing out. Fill it with garden or potting soil. Add a little sand or peat to make the soil less subject to packing and easier for the worms to get around in. Cover the sides with black paper to simulate underground conditions, and keep the habitat in a cool place out of direct sunlight. This will keep the worms from overheating or retreating to the darker parts of the habitat. You can peel the paper back temporarily so students can observe the worms and tunnels. Put a layer of shredded and moistened dead leaves or newspaper on top of the soil. There should be enough organic matter in the soil for the worms to eat for a couple of weeks. Add small amounts of aged tap water as necessary, but don't let it get soggy.

If you want to keep the earthworms for more than a few weeks, stir in more organic matter, such as shredded dead leaves or fruit and vegetable scraps. If you want to keep a permanent worm habitat in your classroom, use red worms (also called red wigglers, manure worms, or fish worms) rather than night crawlers. They are much better adapted to living in worm bins. You can buy commercially made worm bins for around $40 to $100.

Engage

Diary of a Worm Read Aloud and Worm Wonderings

Making Connections: Text to Self

Ask students,

? Have you ever kept a diary or journal?

? What are some things that people include in diaries and journals? (Answers might include: dates, things they did, their ideas, pictures, and drawings.)

Then tell students that you have a very unusual diary to share with them. Show the cover of *Diary of a Worm* and introduce the author and illustrator. Read the book aloud to the class.

QtA (Questioning the Author)

"Questioning the author" (QtA) is an interactive strategy that helps students comprehend what they are reading. When students read in a QtA lesson, they learn to question the ideas presented in the text while they are reading, making them critical thinkers, not just readers. This strategy can be very effective in the science classroom as a way to keep students from acquiring misconceptions from the text or illustrations in a picture book or textbook. After reading *Diary of a Worm,* turn back to some of the dates in the diary and model the QtA strategy.

Say, "I know that Doreen Cronin meant *Diary of a Worm* to be a funny, imaginary story. I realize that worms can't talk or write, and they certainly don't wear clothes! But in the book the author wrote some things about worms that did make me wonder. There were some pictures in the book that I wondered about as well. I wrote down some of these questions, or wonderings that I would like to ask the author." Then share the following questions with the students. You can demonstrate how to interact with the text by placing sticky notes on the corresponding pages with these questions (or

just large question marks) on them.

"March 20: When we dig tunnels, we help take care of the earth."

? The worm in this picture is digging a tunnel. Do worms really make tunnels? Do they push with their heads like the worm in the picture is doing?

"March 20: Never bother Daddy when he's eating the newspaper."

? The worm's "Daddy" has teeth. Do worms have teeth?

"March 29: Today I tried to teach Spider how to dig."

? It looks like the worm in this picture is crawling into the ground. On the next page the worm is crawling above the ground. Can worms crawl above ground?

? The illustrator, Harry Bliss, draws all of the worms in this book with eyes and noses. Do worms have eyes and noses?

"April 10: It rained all night and the ground was soaked. We spent the entire day on the sidewalk."

? I see the sun in this picture. Do worms like to be in the sun? Do worms prefer dry places or damp places?

"June 15: My older sister thinks she's so pretty. I told her that no matter how much time she spends looking in the mirror, her face will always look just like her rear end."

? Is this accurate—a worm's head and tail look just alike? Can you tell a worm's head from its tail?

Next, explain that scientists are always asking questions and making observations about the world around them. Scientists who study animals are called *zoologists.* Tell students that they are going to make some observations, read a nonfiction book, and even do an experiment the way a zoologist would in order to find the answers to these worm wonderings. Give students a worm journal for each, and have them write their names and the date on the cover. Next, read the "Worm

Wonderings" on page 1 of the journal together. Have students make predictions by writing "yes" or "no" for each question, and then turn and talk to a partner about why they made each prediction. (All of these questions will be answered at various points in this lesson through observation, experimentation, or by reading a nonfiction book.)

Then have students brainstorm and share some of their own wonderings about earthworms. They should write their questions on the last two rows of the chart. (If their own questions are not answered by the end of the lesson, encourage them to do further research on their own). These journals and questions will serve to guide the rest of the activities in this lesson.

explore/explain

Earthworm Observations

In advance, prepare a worm bin in a clear plastic container with a ventilated lid. Cover the sides with black paper and keep in a cool, dark place until you begin this activity. Then bring out the worm bin and peel back the paper on the sides to expose the earthworm's tunnels. Allow students to observe the tunnels. Describe how you set up and maintain a worm bin. Then have them turn to page 1 of their journals and fill in their learnings for the first question:

1 Do worms make tunnels? (yes)

Then explain that students will be getting up close and personal with some live earthworms in order to collect *data*. Data is information. Data collected by a zoologist might be in the form of measurements, labeled scientific drawings, or observations of an animal's habitat, body covering, body parts, or behaviors. Discuss that good zoologists always record the data they collect about the animals they are observing. Their data can help them explain what they are observing about the animals.

Before students begin collecting their earth-

worm data, remind them that **all living things should be treated humanely**, including earthworms. Students should handle the earthworms with care and respect and then return them to the safety of the worm bin.

Give each pair of students an earthworm on a paper plate. Instruct them not to touch or handle the worms until they get to journal question #4, because the worms can dry out if handled too much. Have them use hand lenses to make some initial observations for question #1 and then draw and color their earthworm in the box provided on page 2 of the journals. After reading question #2, they can identify and label the head, tail, and saddle.

After reading question #3, they should carefully count as many segments as they can and then circle an estimate (10, 20, 30, 40, 50, or more than 50).

Observing an earthworm

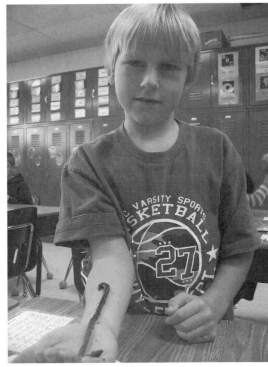

Feeling the bristles

Measuring an earthworm

Night crawlers generally have about 150 segments, so it will be difficult to count them all. It should be sufficient for students to estimate that there are over 50 segments on an earthworm. Before students complete questions #4 through #6, review humane treatment of the earthworms. In #4, they observe a worm's response to touch. In #5, they learn about the bristles of the earthworm, which act like tiny legs to help the earthworm grip and pull itself forward. If they gently place the earthworm on their arm, they may be able to feel the scratchy bristles as the earthworm moves on their skin.

For journal question #6, explain how to measure the length of an earthworm humanely: Supervise as one student holds both ends of the worm and GENTLY stretches it to full length on the paper plate while another student measures in centimeters. **Students should be extremely cautious when they are stretching the worms out to measure their length. Earthworms can break!** Next, have them measure their worms' width and also describe how the worm feels.

Have students compare their observations to others and discuss any differences they find. Then refer to the Worm Wonderings on page 1 in the worm journal and ask students which questions they can now answer after observing the earthworms. Students should be able to answer the following questions:

3 Can worms crawl above ground? (Yes—they can crawl on a paper plate.)

4 Do worms have eyes and noses? (Not that we could see with a hand lens.)

6 Can you tell a worm's head from its tail? (Yes—the head is closest to the thickened part, called the saddle.)

Next, write the word *adaptation* on the board and discuss that an adaptation is a body part or behavior that helps an animal meet its needs. Ask

? What is an earthworm adapted for? (Answers might include: burrowing through, living in, and eating soil)

? What adaptations can you see that might help the worm live in soil? (It has segments to help it bend and move through soil, it can move by grabbing on to the plate and pulling its body forward with its tiny bristles, it is slimy, which might help it move through the soil, it doesn't have eyes because it lives in the dark, it moves away when you touch it, which would help it get away from predators might be among the answers.)

 explore/explain

Damp or Dry? Experiment and A Day in the Life of an Earthworm

Discuss that scientists can't answer all of their questions by observation and measurement alone. Often they need to design experiments to answer questions, such as question #5 on page 1 of the journal, *Do worms prefer damp or dry places?* Tell students that you have prepared an experimental chamber for investigating earthworms. Show them a shoebox with a lid (either a plastic shoebox with the sides covered or a cardboard shoebox). Ask them to think about how you could **humanely** test whether earthworms prefer damp or dry places using the experimental chamber and two earthworms. Then have them turn to page 3 of their worm journals and read the prediction, "I think the earthworms will move to the damp/dry paper towel (circle one)."

Explain that scientists often begin their experiments by making a *prediction,* which is a guess based on what they already know. Ask students to think about what they already know about earthworms and their behaviors before making a prediction. Then have students circle their prediction (damp or dry). You can do the following experiment as a class demonstration. Or you can prepare several experimental chambers and have students work in groups. Next, place a wet paper towel on one side of the chamber and a dry paper

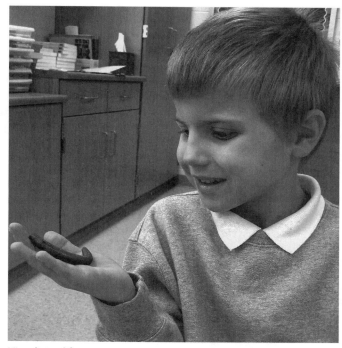

Visualizing life as a worm

towel on the other, with the paper towels about 1 cm apart. Place two earthworms in the space between the paper towels, and close the lid. Wait 10 minutes.

Visualizing

While you are waiting, have students turn to A Day in the Life of an Earthworm on page 4 of their journals. Tell them that they are going to pretend to be a real live earthworm. Not an earthworm that can talk, write a diary, and wear clothes, but an earthworm they might find in a shovelful of dirt in their back yard. Have students close their eyes and visualize what life would be like. What adaptations do they have for living underground? What things do they do that help the soil and the earth? What dangers would they face? What would a day in their life be like? Have them open their eyes, discuss with a neighbor, and then begin writing a realistic story about a real earthworm.

After 10 minutes, remove the cover of the experimental chamber and observe the location of the worms. Have students record the results using a check mark for each worm. Put the worms back in their container, and repeat the experiment using two "fresh" worms. Then have students write a conclusion with evidence based on their data, such as, "When given a choice, worms move to a damp paper towel more often than a dry paper towel." Have students discuss possible reasons. Then have students record their answer to the following question on the Worm Wonderings page of their journal:

5 Do worms prefer damp or dry places? (Worms prefer damp places.)

If you choose to perform this experiment in groups, have students compare their results to others and discuss possible reasons for differences in their data.

explain

Wiggling Worms at Work
Read Aloud

Have students go back to the "Worm Wonderings" on page 1 of their worm journals. Ask

? Were there any wonderings that you couldn't answer completely by observing and measuring earthworms and doing the Damp or Dry? experiment? (Yes—question number 2. We couldn't see if worms have teeth. Students may still have some unanswered questions of their own.)

📖 Determining Importance

Tell students that scientists can't always find out everything they want to know by making observations or doing experiments. They often research using nonfiction books and scientific articles to find the answers to their questions. Sometimes they also look to these sources to see how the results of their experiments compare to what is already known. Explain that you have a nonfiction book called *Wiggling Worms at Work* by Wendy Pfeffer that might help them answer the question about teeth, and clear up some of their own wonderings. They

can also compare what they have observed to what is known about worms by reading this book. The author worked closely with several scientists when she wrote the book to make sure that it contained accurate information about earthworms.

Have students listen carefully for the answers to the "Worm Wonderings" in their journals as you read pages 1–17 aloud (you may want to skip the rest of the book, which is about the earthworm life cycle, as it is not necessary for this lesson). Have students signal by raising their hands or touching their noses when they hear an answer or a verification of a learning. Stop when each question is answered, discuss briefly, and then allow students time to add to the "I leaned" column of their journals. The answers to the worm wonderings can be found on the following pages:

1 Do worms make tunnels? (Page 6—They push aside loose soil and this creates tunnels. Page 8 —Worms tunnel in hard-packed soil by swallowing it.)

2 Do worms have teeth? (Page 10—Worms do not have teeth. They grind the soil in their gizzards.)

3 Can worms crawl above ground? (Page 12— Sometimes worms crawl above ground.)

4 Do worms have eyes and noses? (Page 17— Worms have no eyes, no noses, no ears, and hardly any brain at all.)

5 Do worms prefer damp or dry places? (Page 17—Worms must live in damp soil since they breathe air through their moist skin. In the hot sun their skins dry up and they can't breathe.)

6 Can you tell a worm's head from its tail? (Not found in the book.)

After reading, discuss how the information found in the book compared to the students' predictions and observations. Also discuss if any of the students' own wonderings were answered by the book.

elaborate

How Do Worms Help the Earth?

Rereading

Reread the last page of *Diary of a Worm*:

"August 1: The earth never forgets we're here."

Then ask

? What does the author mean by this? How do worms help the earth?

Next, follow the directions on page 33 of *Wiggling Worms at Work* for setting up the experiment called "Do Earthworms Really Help the Soil?" This can be done at school or as a home project. Students will make long-term observations of two plants to answer the question, "Which plant will grow bigger and better: one planted in regular soil or one planted in soil enriched with worm castings?" Brainstorm a list of controlled variables (things to keep the same to make sure the experiment is fair), such as

- same type of plant,
- same sized plant,
- same sized pot,
- same amount of water,
- same temperature, and
- same amount of light.

Then set up the experiment, and determine how data will be collected (measuring the height of the plants, comparing the number of leaves, and so on). Students can record their data and observations on the back cover of their journals. The plant in the casting-enriched soil should grow bigger and better, if so, students can infer that earthworms really do help the soil.

Cloze Paragraph/Rereading

Next, pass out the How Do Worms Help the Earth? student page. Directions for students are

as follows:

1 Cut out the cards in the boxes below.

2 Read the cloze paragraph, and lay the cards where you think they belong on the blanks.

3 Listen carefully while your teacher reads pages 5–13 of *Wiggling Worms at Work*.

4 Move the cards around if necessary, and glue or tape them on the page.

5 On the back, draw a picture that shows an earthworm helping the earth!

The paragraph should read:

Worms loosen the *soil* as they wiggle along. As worms twist and turn, they create *tunnels*. *Air* flows along these tunnels. *Rainwater* trickles down. Moist ground helps plants grow better. Worms *digest* leaf and plant bits. What's left passes through a worm's body and comes out its tail end in the form of worm *castings*. They make good plant food. Worms help new *plants* begin to grow.

evaluate

Save the Worms Poster and Wiggling Worms Quiz

Tell students that their study of earthworms can help save the worms! Pretend that there is a debate going on concerning a local golf course. The golf course has been overrun with moles who are digging up the grass in search of their favorite food source: earthworms. The groundskeeper, Mr. Spackler, wants to spray pesticides on the ground to kill the earthworms. He thinks that killing their food source will make the moles go away. He also thinks that earthworms are disgusting, worthless creatures, and doesn't want them on his golf course.

Pass out the Save the Worms Poster grading rubric. Tell students that their assignment is to create a poster to convince Mr. Spackler that killing the earthworms is a bad idea. Their posters should promote earthworms as fascinating, useful

Sample "Save the Worms" poster

animals that can actually help the grass on the golf course grow better. Have students or groups of students create their "Save the Worms" posters using the information from their worm journals. They may want to do more research on the internet or through nonfiction reading in order to support their arguments (see Websites and More Books to Read). Posters should include

4 Points: Descriptions of four different earthworm adaptations

3 Points: A labeled, detailed, full-color drawing of an earthworm

2 Points: Two ways earthworms help the earth

1 Point: One additional fascinating fact about earthworms

Extra Credit: A poem, song, rap, or cheer about saving the worms.

You can use the rubric to score completed posters and make comments. As an additional evaluation,

"Save the Worms" poster session

you may want to give the Wiggling Worms quiz. Answers are as follows:

1 head

2 saddle

3 segment

4 tail

5 castings

6 c

7 Answers might include two of the following: Use damp soil; add leaves, compost, or other decaying matter; add newspaper; keep in a cool, dark place; keep out of the sun; and treat worms humanely.

8 Answers might include two of the following: same type of plant, same-sized plant, same-sized pot, same amount of water, same temperature, and same amount of light.

9 Answers might include two of the following: Worms loosen the soil by creating tunnels, worm tunnels let air and/or water into the soil, and worms make castings which make good plant food.

Inquiry Place

Have students brainstorm testable questions about worms, such as

? Does a worm move headfirst or tail first more often?

? Do worms prefer different types of food?

? Do worms react to strong smells?

? Do worms prefer light or dark?

? How long does it take a worm to burrow into the soil?

? Do different types of earthworms have a different number of segments?

Then have students select a question to investigate as a class, or have groups of students vote on the question they want to investigate as a team. After they make predictions, have them design an experiment to test their predictions. Students can present their findings at a poster session or gallery walk.

More Books to Read

Cronin, D. *Diary of a spider.* 2005. New York: Joanna Cotler Books.

Summary: Fans of *Diary of a Worm* will enjoy reading the humorous adventures of young worm's friend, Spider.

Himmelman, J. 2001. *An earthworm's life.* New York: Children's Press.

Summary: Simple text and illustrations describe the daily activities and life cycle of the earthworm using an engaging story format.

Kalman, B. 2004. *The life cycle of an earthworm.* New York: Crabtree Publishing.

Summary: This book details characteristics of earthworms and their important role in nature. Fairly large text, simple explanations, colorful photographs, table of contents, glossary, and index make this a great choice for budding researchers. The common misconception that earthworms cut in half can regenerate major body parts is dispelled.

Oxlade, C. *Soil (Materials, materials, materials).* Chicago: Heinemann Library.

Summary: A simple presentation of information about soil, including its composition, properties, and some of its uses. Includes large full-color photographs, bold-print words, fact file, glossary, more books to read, and index.

Rosinsky, N. 2003. *Dirt: The scoop on soil*. Minneapolis: Picture Window Books.

Summary: Simple text and colorful illustrations explain what soil is made of, how worms and other burrowing creatures help soil, and why dirt is important to all life on Earth. Includes a table of contents, glossary, simple experiments, and a FactHound website with links to other safe, fun websites on the topic.

Tomecek, S. 2002. *Dirt: Jump into science*. Washington, DC: National Geographic Children's Books.

Summary: Young readers get down in the dirt with a friendly star-nosed mole as he shows off the different parts of his garden and some amazing creatures who live in the dirt. During this exciting tour, children discover what soil is, how it is formed, and why this substance is vital to plants, animals, and humans. Colorful art and simple text show how the soil that plants grow in differs from the soil that building foundations sit in, and help youngsters understand why this precious resource must be conserved.

Websites

Visit Discovery Kids Worm World for all the dirt on worms. *http://yucky.kids.discovery.com/flash/worm/index.html*

Visit "The Adventures of Herman" to learn all about Squirmin' Herman the Worm (also available in Spanish) *www.urbanext.uiuc.edu/worms/index.html*

My Worm Journal

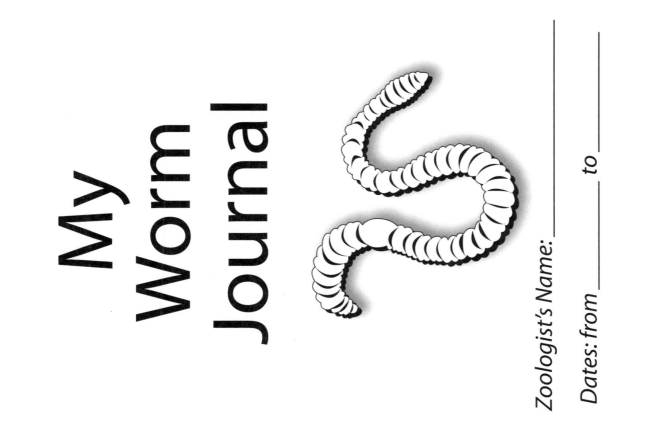

Zoologist's Name: _____

Dates: from _____ to _____

Worm Wonderings

I wonder...	I predict...	I learned...
1. Do worms make tunnels?		
2. Do worms have teeth?		
3. Can worms crawl above ground?		
4. Do worms have eyes and noses?		
5. Do worms prefer damp or dry places?		
6. Can you tell a worm's head from its tail?		

My Own Worm Wonderings

I wonder...	I predict...	I learned...

A Day in the Life of an Earthworm

Pretend you are a real earthworm. What are some of your adaptations for living underground? How do you help the earth? What dangers would you face? What would a day in your life be like?

Damp or Dry? Experiment

Question: Do earthworms prefer damp or dry places?

Prediction: I think the earthworms will move to the **damp/dry** paper towel (circle one).

Procedure:

1. Put two worms in the box and cover.
2. Wait 10 minutes.
3. Observe the worms. Record the results using a ✓ for each worm.
4. Repeat with two new worms.

Worm's Response	Trial #1	Trial #2
Damp Paper Towel		
Dry Paper Towel		

Conclusion: Do earthworms prefer damp or dry places? What is your evidence?

_____

_____

_____

Earthworm Observations

1. Draw and color your earthworm in the box below.

2. An earthworm's **head** is closest to the thickened part of its body, called the **saddle.** Label your earthworm's head, tail, and saddle in the box above.

3. An earthworm's body is made up of rings, or **segments,** that let it bend. How many segments can you count?   10   20   30   40   50   >50

4. What does the worm do when you gently touch it?

   _____

5. An earthworm has eight tiny **bristles** under each segment that act a little like legs. They help it move. Gently place the worm on your arm. Can you feel the bristles? _____

6. Data Table

Length	Width	Feels Like
cm	cm	

Name: _____

How Do Worms Help the Earth?

Directions:

1. Cut out the cards in the boxes below.

2. Read the cloze paragraph, and lay the cards where you think they belong on the blanks.

3. Listen carefully while your teacher reads pages 5–13 of *Wiggling Worms at Work*.

4. Move the cards if necessary, and glue or tape them on the page.

5. On the back, draw a picture that shows a worm helping the earth!

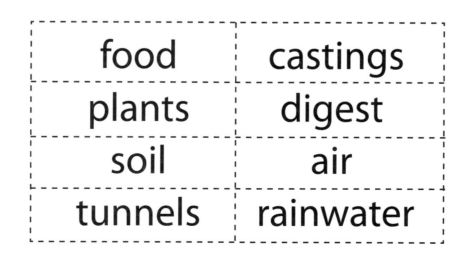

food	castings
plants	digest
soil	air
tunnels	rainwater

How Do Worms Help the Earth? *cont.*

Worms loosen the _____ as they

wiggle along. As worms twist and turn, they create

_____. _____ flows

along these tunnels. _____ trickles

down. Worms _____ leaf and plant

bits. What's left passes through a worm's body and

comes out its tail end in the form of worm

_____. They make good plant

_____. Worms help new

_____ begin to grow.

Name: _____

Save the Worms

Poster Rubric

Name(s): _____

Your poster includes:

4 Points: Descriptions of four different earthworm adaptations

        4   3   2   1   0

3 Points: A labeled, detailed, full-color drawing of an earthworm

                3   2   1   0

2 Points: Two ways earthworms help the earth

                        2   1   0

1 Point: One additional fascinating fact about earthworms

                                1   0

Extra Credit: Your poster includes a poem, song, rap, or cheer about saving the worms.

                                1   0

Total Points_____/10

Comments: _____

Name: _____

Wiggling Worms
Quiz

Label the earthworm picture below. Use the WORD BANK.

WORD BANK	
castings	saddle
tail	segment
head	

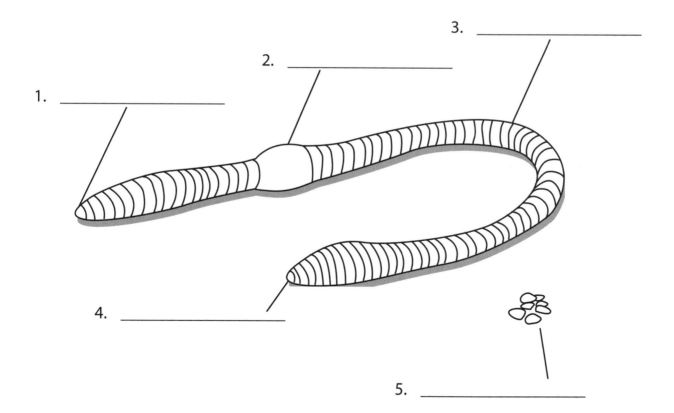

1. _____

2. _____

3. _____

4. _____

5. _____

Wiggling Worms Quiz cont.

6. How do an earthworm's tiny bristles help it live?

    a. The earthworm uses its bristles to breathe.

    b. The earthworm uses its bristles like tiny roots to soak up water.

    c. The earthworm fastens its bristles to the soil, helping it to move.

7. Marcus wants to raise earthworms in the classroom. Describe two things he should keep in mind when setting up the worm bin.

    a._____

    b._____

8. Wendy wants to set up an experiment to see if plants grow better in soil with or without earthworms. What are two things Wendy should keep the same in her experiment so that it is a fair test?

    a._____

    b._____

9. Describe **two** ways earthworms help the soil:

    a._____

    b._____

Over in the Ocean

Description

Learners make observations and ask questions about coral reefs and the animals that inhabit them. They create and use categories to sort ocean animal models based upon their characteristics and then create a class question book about coral reefs.

Suggested Grade Levels: K–2

Lesson Objectives Connecting to the Standards

Content Standard A: Science as Inquiry
- Ask a question about objects, organisms, and events in the environment.

Content Standard C: Life Sciences
- Understand that distinct environments support different types of organisms.

Featured Picture Books

Title	*Over in the Ocean: In a Coral Reef*	*Coral Reef Animals (Animals in Their Habitats Series)*
Author	Marianne Berkes	Francine Galko
Illustrator	Jeanette Canyon	
Publisher	Dawn Publications	Heinemann
Year	2004	2003
Genre	Story	Non-narrative Information
Summary	Based on the traditional song "Over in the Meadow," this ocean-animal counting book features vibrant polymer clay illustrations.	Describes coral reefs, where they can be found, the animals that inhabit them, and how to protect them.

Time Needed

This lesson will take several class periods. Suggested scheduling is as follows:

Day 1: **Engage** with *Over in the Ocean: In a Coral Reef* read aloud.

Day 2: **Explore/Explain** with Ocean Animal Sorting.

Day 3: **Elaborate** with *Coral Reef Animals* read aloud and Questioning.

Day 4+: **Evaluate** with Coral Reef Question Book.

Materials

Coral Reef Toobs (1 Toob contains enough animals for 2 pairs of students, or use other lifelike ocean animal models or laminated color photographs of coral reef animals.)

Sticky notes (1 pack per student)

A variety of nonfiction books and magazines about coral reefs (see More Books to Read)

Coral Reef Toobs are available from
www.boondoggles.com
www.christianbook.com
www.fatbraintoys.com
www.playfairtoys.com

Student Pages

Ocean Animal Sorting (1 per pair or group of 3–4 students)

Coral Reef Questions

Background

Young children have an innate curiosity about the natural world—animals, plants, rocks, water, clouds, Sun, and so on. The National Science Education Standards recommend that children in grades K–4 have opportunities to observe the objects, organisms, and events in the environment and ask questions that they can answer using reliable sources of scientific information combined with their own observations. Questions arise from the sense of wonder and the natural interests of children as they experience the world around them. This lesson uses colorful coral reef animal models to engage children in asking questions, making observations, describing the different structures of animals, and sorting them into groups based upon their characteristics. Student questions about coral reefs are compiled into a class book, and they are encouraged to use a variety of scientific books and resources to answer them.

Over in the Ocean: In a Coral Reef presents the incredible biodiversity within a coral reef. Ten fascinating sea animals, including *invertebrates* (such as octopuses) and *vertebrates* (such as dolphins and fish) are introduced in this counting book, written in the style of "Over in the Meadow." Students will enjoy manipulating lifelike models of some of the animals featured in the book, describing them, and sorting them into groups. More information about each animal can be found on pages 28 and 29 of the book. The author's note on page 27 discusses how many babies coral reef animals actually have (quite a few, with the exception of dolphins, which have only one or two babies at a time). In the book *Coral Reef Animals,* students will be interested to learn that coral reefs are actually composed of the bodies of tiny animals called *coral polyps.* Coral polyps belong to a group of invertebrates with soft, jellylike bodies. They attach themselves to the seafloor and produce hard skeletons using the calcium in seawater. Reefs form slowly, over thousands of years, as generations of polyps live, build, and die. Coral polyps need sunlight and warm, shallow, clean ocean water, so reefs form only in certain places

around the world. Unfortunately, coral reefs are very fragile ecosystems that are being threatened by pollution and human activity. To learn about 25 things your students (and their families) can do to protect coral reefs, go to *www.publicaffairs.noaa.gov/25list.html.*

engage

Over in the Ocean: In a Coral Reef Read Aloud

Inferring

Read the title, and introduce the author and illustrator of *Over in the Ocean: In a Coral Reef.* Show the cover of the book. Ask

? Have you ever heard of a coral reef?

? From looking at the cover of the book, what do you think might be in a coral reef? (Seahorses, fish, and plants are examples.)

Point out that the beautiful artwork in the book was all done with a type of clay. Explain that the illustrator had to make very careful observations of the animals featured in the story by studying them in nature. She actually spent some time swimming near a coral reef to get up close and personal with some amazing ocean animals. Show students the picture of Jeanette Canyon in snorkel gear in the back of the book.

Tell students that this book will help them learn about some of the amazing animals that live in a coral reef. Then read the book aloud. To increase student interaction with the book, invite students to show with their fingers and chorally respond with the number of baby animals pictured on each two-page spread page as you read or sing the verse (for example: "Over in the ocean far away from the sun lived a mother octopus and her octopus **ONE**.")

Rereading

Tell students that you are going to reread the book for a scientific purpose. Explain that you would like them to look very carefully at the different animals that live in the coral reef and observe their special characteristics. Model this by reading the first two-page spread and then describing the octopus. For example, "The baby octopus on this page has eight arms. It has an oval-shaped head. It has two large eyes. It has many pink and white suckers along each arm. It is squirting black stuff into the water." Then reread each page and invite students to make observations about the characteristics of each animal.

Point out that the illustrator was very clever in creating the unique characteristics of each animal. For example, she used a plastic mesh bag that once held tomatoes to create the tiny fish scales on the clownfish. Have students speculate how she may have created other details, such as the thin ribbons of seaweed on some of the pages (she actually used a pasta maker).

explore/explain

Ocean Animal Sorting

Next, tell students that they will be practicing their observation skills using ocean animal models. Pass out a variety of plastic models, one to each student. (Students may want to know the names of the animals. The Coral Reef Toobs have the name of each animal stamped on the bottom.) Ask questions such as

? What color is your animal?

? What patterns does it have on its body?

? How many arms or fins does it have?

? What shape is it?

? Does your model look like any of the animals from the book? Which animal?

Next, pass out additional sea animal models so that each pair of students has a set of six or more

Modeling Ocean Animal Sorting

animals to work with. Then pass out the Ocean Animal Sorting student page to each pair. Explain that they will be using the circles on the page to sort their animals into two groups. Model how to do this, for example: "These animals all have blue on them. These animals do not. I can sort them into two groups: Blue and not blue."

Then model another way to sort the animals based upon a different characteristic, for example: "These animals all have fins. These animals all have arms. I can sort them into two groups: animals with fins and animals with arms."

Have students practice sorting their animals into two groups based on one characteristic. They should be able to explain to you what characteristic they used to sort the animals. Then have pairs switch places with other pairs and guess what characteristic was used to sort the animals.

After students have had the opportunity to sort several ways, ask

? What different characteristics did you use to sort your animals? (Number of fins, number of arms, color, stripes, spots, and size are possible examples.)

? Were there animals that were hard to sort? Why?

? What questions do you have about the animals you have been sorting?

elaborate

Coral Reef Animals
Read-Aloud and Questioning

Tell students that reading the book *Over in the Ocean: In a Coral Reef* and sorting the ocean animal models made you want to learn more about coral reefs. Ask

? Where can I look to get more information about coral reefs? (Possible answers include: in a book, in a magazine, and on the internet.)

? What kind of book could I read to get true facts and information about coral reefs? A fiction or a nonfiction book? (nonfiction)

📖 Questioning

In advance, select several questions that reading the book *Coral Reef Animals* could inspire, and write them down on sticky notes. (You may substitute any nonfiction book about coral reefs for this activity.) Tell students that you have found a nonfiction book with lots of information about coral reefs. Show students the cover of the book, *Coral Reef Animals*. As you read the book aloud, model your questions (shown in bold below) and then place your sticky notes onto the appropriate pages of the book. You may also want to demonstrate some of the features of nonfiction as you read. For example,

? (Before reading) I have some questions before I even start reading: What is a coral reef? How is it made?

Sorted by stripes/No stripes

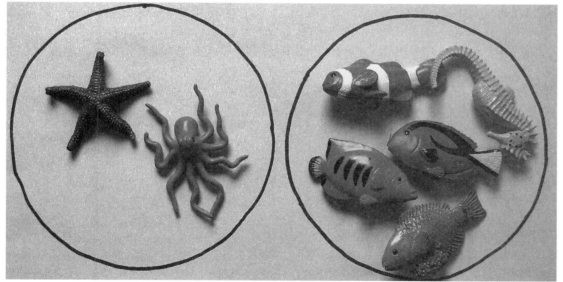

Sorted by arms/no arms

? (p. 4) What is a habitat? (I'll use the glossary in the back of the book to find out. The glossary says a habitat is the place where an animal lives.)

? (p. 5) How big can coral reefs get?

? (p. 7) Are there any coral reefs in the United States?

? (p. 8) How big is a reef squid? (I'll use the inset on this page. It looks like it is smaller than a soccer ball.) I wonder: How big is the largest squid?

? (p. 14) Is a sponge a plant or an animal?

? (p. 19) What is a predator? (I'll use the glossary to find out. The glossary says a predator is an animal that hunts and eats other animals.)

? (p. 23) Is a stingray a kind of fish?

? (After reading) What else can we do to help protect coral reefs? How can I find out more about coral reef animals?

After reading, ask

? What questions did I ask before I even opened the book? (What is a coral reef? How is it made?)

Sorting ocean animals

❓ Did any of my questions lead to other questions? (Yes, the question about the reef squid made you wonder about other squids.)

❓ What features of the book helped me find some of my answers? (glossary, inset)

❓ Were any of my questions answered later in the book? (Yes, we found out that a sponge is an animal.)

❓ Did I ask any questions after I finished the book? (yes)

Explain that thoughtful readers ask questions before, during, and after they read. Go

Writing questions on sticky notes

back through the book, and have students share questions they had as you were reading, or questions that they had after you finished the book. Show that you value students' questions—repeat them as they are asked, ponder them a bit, and really celebrate student thinking. Talk about how hearing others' questions can inspire new ones of your own.

evaluate

Coral Reef Question Book

Next, discuss that questioning is very important in science. Questions help lead scientists to answers about the world. Many problems have been solved and new things invented or discovered because scientists had questions. Scientists don't always find the answers to all of their questions, but they ask a lot of questions anyway. Tell students that the whole class is going to work together as scientists to create a Coral Reef Question Book about coral reef animals.

In advance, collect a variety of nonfiction books and magazines on coral reef animals (see More Books to Read for some suggestions). Make sure you have enough resources so that every student or pair of students can browse at the same time (or have students take turns visiting a station containing the resources). Give each student a sticky-note pad. Invite students to read silently or in pairs, generating questions on sticky notes as they read. They can place the sticky notes directly on the pages of each book.

Encourage students to read more than one book or article. After reading time, bring them back together and have them talk about some of the books and articles they read and the questions they wrote. Discuss whether any of their questions were inspired by reading the questions on other students' sticky notes. You may want to spend several class periods on this questioning activity.

Then pass out the Coral Reef Questions student page. Have students choose an animal

from one of the read alouds or the additional resources, draw a colorful picture of it, and then write two or three interesting questions about the animal. Collect all of the student pages, and bind or staple them together in a book.

As students do more reading about coral reef animals, they may discover answers to some of the questions in their class book. Encourage them to write answers, as well as sources (name of book and author), on the backs of the book pages.

Inquiry Place

Have students brainstorm researchable questions about coral reef animals, such as

? How many babies do the animals in *Over in the Ocean: In a Coral Reef* really have?

? How many fins do fish have? Do different kinds of fish have different numbers of fins?

? What is the largest fish in the ocean? The smallest?

? Where on earth are coral reefs?

Students can choose a question to investigate in teams or as a class and add them to their class Coral Reef Question Book.

More Books to Read

Berger, M. *Life in the sea*. 1994. New York: Newbridge Educational Publishing.

Summary: Full-color photographs, captions, and spare text describe the types of life that can be found at different levels of the sea, from seahorses near the top to giant clams at the bottom.

Cole, J. 1998. *Magic School Bus takes a dive: A book about coral reefs*. New York: Scholastic.

Summary: With help from Red Beard's treasure map, Ms. Frizzle and her class discover the animals and plants of the coral reef.

Earle, S. 2001. *Hello fish! Visiting the coral reef*. Washington, DC: National Geographic Children's Publishing.

Summary: Up-close, full-color underwater photography and brief text introduce 12 fish that live in and around coral reefs.

Pallotta, J. 1991. *The underwater alphabet book*. Boston: Charlesbridge.

Summary: Vivid illustrations and fact-filled, entertaining text take the reader on an A to Z journey through a coral reef.

Pluckrose, H. 1995. *Sorting (Math Counts Series)*. Chicago: Childrens Press.

Summary: Full-color photographs and simple text demonstrate various ways everyday objects can be sorted and reasons why people sort things.

Pratt, K. J. 1994. *A swim through the sea*. Nevada City, CA: Dawn Publications.

Summary: An A to Z introduction to various forms of ocean life, as seen through the eyes of Seamore the seahorse.

Pulley Sayre, A., and J. 2006. *One is a snail, ten is a crab: A counting by feet book*. Cambridge, MA: Candlewick Press.

Summary: If one is a snail and two is a person, we must be counting by feet! Just follow the sign to the beach, where crabs, dogs, insects, and snails offer their feet for counting in a number of silly, surprising combinations—from one to one hundred.

Telford, D., and R. Theodorou. 2006. *Inside a coral reef*. Chicago: Heinemann.

Summary: Describes life in the Great Barrier Reef and the effects human actions have on coral reefs.

Whitehouse, P. 2003. *Hiding in a coral reef*. Chicago: Heinemann.

Summary: Learn how animals living in a coral reef use various types of camouflage to survive, capture prey, or hide from predators.

Websites

National Oceanic and Atmospheric Administration's 25 Things You Can Do to Save Coral Reefs
www.publicaffairs.noaa.gov/25list.html

Enchanted Learning Coral Reef Animals
www.enchantedlearning.com/painting/coralreef.html

Ocean Animal Sorting

Directions: Sort your ocean animals into two groups. Have someone guess how you sorted them!

Name: _____

Coral Reef Questions

Name of Animal:_____

_____

My questions about this animal: _____

_____

_____

Be a Friend to Trees

Description

Learners explore the variety of products made from trees, the importance of trees as sources of food, shelter, and oxygen for people and animals, and ways to conserve trees.

Suggested Grade Levels: K–4

Lesson Objectives Connecting to the Standards

Content Standard A:
Science as Inquiry:
Abilities Necessary To Do
Scientific Inquiry

- Ask a question about objects, organisms, and events in the environment.

Content Standard B:
Life Science: Organisms and Their Environments

- Understand that all animals depend on plants. Some animals eat plants for food. Other animals eat animals that eat the plants.

Content Standard F:
Science in Personal and Social Perspectives: Types of Resources

- Understand that the supply of many resources is limited. Resources can be extended through recycling and decreased use.

Featured Picture Books

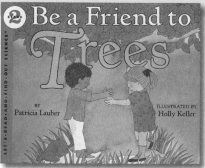

Title	*Our Tree Named Steve*	*Be a Friend to Trees*
Author	Alan Zweibel	Patricia Lauber
Illustrator	David Catrow	Holly Keller
Publisher	G. P. Putnam's Sons	HarperTrophy
Year	2005	1994
Genre	Story	Non-narrative Information
Summary	In a letter to his children that is both humorous and poignant, a father recounts memories of the role that Steve, the tree in their front yard, has played in their lives.	Discusses the importance of trees as sources of food, oxygen, and other essential things, and gives helpful tips for conserving this important natural resource.

Time Needed

This lesson will take several class periods. Suggested scheduling is as follows:

Day 1: **Engage** with *Our Tree Named Steve* read aloud.

Day 2: **Explore/Explain** with Sorting Chart and *Be a Friend to Trees* read aloud.

Day 3: **Elaborate** with My Favorite Tree.

Day 4: **Evaluate** with Be a Friend to Trees Poster.

Materials Per Group of 3–5 Students

Sorting chart made from chart paper with a large Venn diagram drawn on it

Boxes or bins, 1 per group, filled with several of the following tree parts or products (actual objects or pictures of the objects) described in the book *Be a Friend to Trees*:

From Trees:

Wooden block

Writing or construction paper

Newspaper

Small cardboard box or paper milk carton

Apple, orange, pear, cherry, or peach

Walnut, almond, pecan, or hazelnut in the shell (Check to see if you have students with tree nut allergies in your class, and use only pictures if you do.)

Small tree branch with leaves

Pine needles

Piece of tree bark

Paper towel

Paper grocery bag

Sealed baggie or balloon blown up with air and marked "Oxygen" (This will represent oxygen although it also contains other gases.)

Also include some of the following objects (actual objects or pictures):

Not From Trees:

Plastic objects (such as small toys, markers, balls, and containers)

Metal objects (such as keys, foil, and spoons)

Glass marble

Rock

Small pumpkin, squash, carrot, or potato

Cotton, polyester, or nylon cloth

Reusable net or canvas grocery bag

Sealed baggie blown up with air and marked "Carbon Dioxide" (This will represent carbon dioxide although it contains other gases.)

Both:

Pencil with eraser

Plastic bottle of maple syrup

Chocolate in a foil wrapper

Materials Per Student for My Favorite Tree Activity

Crayon with the paper removed

Pencil

Clipboard (or notebook to use as a writing surface)

Student Pages

My Favorite Tree journal (4 single-sided pages stapled together)

Be a Friend to Trees Poster rubric

Background for Teachers

Trees are one of Earth's most important natural resources. We depend on trees for food and wood products, water and soil conservation, shade, beauty, and, most important, the oxygen they add to the air. It is essential for students to understand and appreciate the importance of trees to humans and all life on Earth, and to realize that their actions can have an impact on trees. The National Science Education Standards state that children should have a variety of experiences to help them understand how animals, including humans, depend on plants, and that the supply of natural resources like trees can be extended through decreased use and recycling. The Explore/Explain phase of this lesson allows students to explore our dependence on trees by observing and sorting various products that come from them. After reading about how humans and other animals depend on trees, they also learn a variety of ways that they can be a friend to trees.

Nurturing a sense of wonder about trees will encourage students to do more to protect and conserve this vital resource. The Standards suggest that children in the elementary grades build understandings of biological concepts through asking questions and through direct experience with living things, their life cycles, and their habitats. The Elaborate phase of this lesson involves students observing trees closely, noticing details in their shapes, leaves, and bark, and generating wonderings about their favorite trees.

engage
Our Tree Named Steve
Read Aloud

📖 Stop and Jot, Turn and Talk

Before reading the book, *Our Tree Named Steve*, engage students by saying, "From where you are sitting, look around and think of everything in this room that might be different if there was no such thing as a tree." Allow some quiet thinking time, and then have students turn and talk to a neighbor. You may want to have students stop and jot their ideas before sharing with a neighbor.

Inferring

Explain that you have a book to share about a very special tree. Show the cover of *Our Tree Named Steve*, and then introduce the author and illustrator. Ask

? What are you thinking this story is about? Why do you think so?

Synthesizing

Read the book aloud, stopping after page 5 ("… Mom and I got the hint and asked the builder to please save Steve.") Then ask

? Now what are you thinking this story is about?

You may want to stop at key points in the story to allow students to discuss their thinking about the story's meaning.

A Tree Named Steve graphic organizer

Questioning

After reading, draw a large tree trunk on the board or chart paper and label it "Steve." Ask

? How did the tree get its name? (The youngest daughter couldn't pronounce "tree" and called it "Steve.")

You can write the students' responses to the following two questions as "branches" of the tree. Ask

? How did the family use this special tree when it was alive? (Answers can include: as a swing holder, target, third base, hiding place, jump-rope turner, clothesline, hammock-holder, and sewer-water remover.)

? How did the family use the tree after it blew over in the storm? (They used the wood to build a tree house.)

Then ask

? How did the tree protect the family "to the very end"? (It didn't fall on their house, the swings, the dog's house, or the garden.)

Synthesizing

? Now what are you thinking the story is about?

? How does the story make you feel?

? Have you ever had a special tree? What made it special?

? What are some ways that trees help us?

explore/explain

Sorting Chart and *Be a Friend to Trees* Read Aloud

In advance, create boxes or bins filled with an assortment of items that came from trees, items that did not come from trees, and items that contain both wood products and other materials (see materials list). Explain that students will be learning about some of the ways that trees help us by doing a sorting activity and then reading

Sorting objects

a nonfiction book. Divide students into groups, and distribute to each group a bin and a Sorting Chart made from chart paper with a Venn diagram (two large intersecting circles) drawn on it. Have students label one circle "From Trees," the other circle "Not From Trees," and the intersection of the circles "Both."

Then have groups observe each object carefully, discuss whether or not they think it came from trees and why, and place it in the appropriate circle on the chart. If they are not sure about how an object should be grouped, they can leave it in the bin for now.

Invite students to justify how they sorted the objects. Ask

? What are some of the objects you think came from trees? Why do you think so?

? What are some of the objects you think did not come from trees? Why do you think so?

? Were there any objects you were unsure about? Why?

Inferring

Next, show students the cover of the book, *Be a Friend to Trees*. Ask

? What do you think this book might be about? Why do you think so?

Determining Importance

Tell students that *Be a Friend to Trees* is a nonfiction book that might help them learn which of the objects came from trees. Introduce the author and illustrator of the book, and then explain that, as you read, you want them to listen for any of the items they placed in the "From Trees" circle on their sorting charts. Ask them to signal (raise hand, touch nose, or in some other way) when they hear about one of the objects.

Questioning

As you read aloud, stop periodically to question students to check for understanding and build interest. Some suggested questions are

? (p. 10) Look at the diagram. What is the first thing that happens in order to make paper? (Wood chips are cooked with chemicals.)

? (p. 10) What are the wood chips called after they become soggy? (pulp)

? (p. 10) What must be done to the pulp after the water is drained off? (It is dried, flattened, and then rolled into paper.)

? (p. 12) What are the only living things that can make their own food? (green plants)

? (pp. 14–20) What are some of the ways that animals use trees? Turn to a neighbor, and share an example from the book. (Possible answers include: Many animals eat leaves, bark, buds, and twigs; squirrels and chipmunks gather nuts to eat; bees collect pollen and nectar; birds roost and nest in trees; and deer hide beneath trees.)

? (p. 21) How do trees help the soil? (They keep it from washing away.)

? (p. 22) What would happen to people and animals if there were no trees or green plants? (There would be no oxygen in the air, and we couldn't breathe.)

? (p. 23) Where do trees make food? (in their leaves)

? (p. 24) What three things do leaves need to make food? (water, carbon dioxide, and sunlight)

? (pp. 30–32) What are some things you can do to be a friend to trees? Turn to a neighbor, and share an example from the book. (Answers might include: Use less paper; re-use paper bags; write on both sides of paper; recycle newspaper; and plant a tree.)

After reading, give students the opportunity to return to their Sorting Charts and move any of the objects to a different spot on the chart if necessary. Then discuss what kinds of things come from trees (such as oxygen, fruits, nuts, and wood and paper products) and what kinds of things don't (such as carbon dioxide, vegetables, plastics, metals, cloth, glass, and rocks).

elaborate

My Favorite Tree

This activity can be done on school grounds, during a field trip to a park or other wooded area, or as a take-home assignment. Take students outside to look closely at a tree. They will each need a copy of the My Favorite Tree journal, a clipboard, a pencil, and a crayon with the paper removed. First, model how to sketch a tree's shape and make careful observations of its leaves and bark.

Observing a leaf

Then, show students how to do a *leaf rubbing:*

1 Find a fallen leaf that is still soft, and place it on your clipboard with the rough or vein side up.

2 Place the journal page over the leaf.

3 Gently rub the long side of the crayon over the leaf.

Next, demonstrate how to do a *bark rubbing:*

1 Pick the part of the bark that you want to make a rubbing of.

2 Place the journal page over that part.

3 Gently rub the long side of the crayon over the bark.

Making a leaf rubbing

Making a bark rubbing

Next, model some of your own wonderings about the tree. (For example: How old is this tree? I wonder who planted it. I wonder if an animal lives in this hollow part. What kind of tree is it?)

Finally, share your thoughts and feelings about the tree by explaining why you chose the tree for your journal. (For example: This is my favorite tree because the bark peels up in places and looks like paper. I like how I can fit my arms all the way around the trunk. I have never seen a tree like it before. I feel peaceful when I sit with my back leaning against the trunk.)

If this activity is to be done at home, students can take their journals home and complete them with an adult helper. If this activity is to be done at school or on a field trip, allow students to look at several trees before deciding on a favorite to include in their journals. When you return to the classroom, have students share their journals with each other.

evaluate

Be a Friend to Trees Poster

Ask

? What does it mean to "be a friend to trees?" (to do things that will help protect or conserve trees)

? Why is it important to "be a friend to trees?" (Answers could include: Trees help animals, humans, and the environment in many ways.)

Pass out the Be a Friend to Trees Poster grading rubric. Have students create a 3-2-1 poster summarizing what they have learned about trees and their conservation. Posters should include

● thorough descriptions of three ways trees are helpful to humans, animals, and the environment,

● two interesting facts about trees, and

● one labeled drawing showing a child being a friend to trees.

For fun and extra credit, students can include their own additional research on trees, or a poem, song, rap, or cheer about being a friend to trees. You can use the rubric to score completed posters and make comments.

Inquiry Place

Have students brainstorm testable or researchable questions such as:

? How can trees be identified?

? How many different kinds of trees are in the schoolyard?

? What will happen to a leaf on a tree if it is covered with paper for a length of time?

? What is the world's oldest/tallest/thickest tree?

? How is paper made?

? How can we make recycled paper in the classroom?

Then have students select a question to investigate or research as a class, or have groups of students vote on the question they want to investigate as a team. After they make predictions, have them design an experiment or do research to test their predictions. Students can present their findings at a poster session or gallery walk.

More Books to Read

Gackenbach, D. 1992. *Mighty tree.* New York: Voyager Books.

Summary: Three seeds grow into three beautiful trees, each of which serves a different function in nature and for people.

Gibbons, G. 1984. *The seasons of Arnold's apple tree.* New York: Voyager Books.

Summary: As the seasons pass, Arnold enjoys a variety of activities as a result of his apple tree. Includes a recipe for apple pie and a description of how an apple cider press works.

Mora, P. 1994. *Pablo's tree.* New York: Simon & Schuster Books for Young Readers.

Summary: Every year, Pablo's grandfather decorates a special tree for his birthday.

Shetterly, S.H. 1999. *Shelterwood.* Gardiner, ME: Tilbury House.

Summary: While staying with her grandfather who is a logger, Sophie learns about different kinds of trees, what they need to thrive and grow, and how the bigger trees provide shelter for the smaller ones. Her grandfather teaches her that, when harvesting trees, it is important to let the tallest ones stay to drop their seeds and start a new generation. Sophie discovers that, when we take care of the woods, it provides for us for generations to come.

Silverstein, S. 1997. *The giving tree.* New York: Scholastic.

Summary: Shel Silverstein's poignant story of a boy and a special tree that gives him many things throughout his life.

Udry, J. 1956. *A tree is nice.* New York: HarperCollins.

Summary: This Caldecott award–winning book speaks simply and elegantly of the many pleasures a tree provides.

Worth, B. 2006. *I can name 50 trees today!: All about trees.* New York: Random House.

Summary: While stopping to admire some of the world's most amazing trees, the Cat in the Hat and friends teach beginning readers how to identify tree species from the shape of their crowns, leaves, lobes, seeds, bark, and fruit. Dr. Seuss–inspired cartoons and verses teach readers about many of the trees common to North America.

Websites
The National Arbor Day Foundation
www.arborday.org

Trees for Life
www.treesforlife.org

My Favorite Tree

*By*_____

The Shape of My Favorite Tree

My Favorite Tree cont.

Leaf Rubbing From My Favorite Tree

Observations of the leaf: _____

_____

_____

_____

_____

_____

Bark Rubbing From My Favorite Tree

Observations of the bark:_____

_____

_____

_____

_____

_____

My Favorite Tree cont.

Wonderings about my favorite tree: _____

_____

_____

_____

_____

_____

_____

_____

_____

_____

_____

Why this is my favorite tree: _____

_____

_____

_____

_____

_____

_____

_____

_____

_____

_____

Be a Friend to Trees

3-2-1 Poster Rubric

Name _____

Your poster includes:

3 ways trees are helpful to humans, animals, and the environment.

         1   2   3

2 interesting facts you learned about trees.

         1   2

1 labeled drawing of yourself "being a friend to trees."

         1

For fun and extra credit, you included your own additional research on trees, or a poem, song, rap, or cheer about being a friend to trees.

Total Points_____/6

Comments: _____

That Magnetic Dog

Description

Learners "go fishing" with magnets and discover through exploration that not all metals are magnetic. They verify their findings by reading a nonfiction book about magnets, communicate their findings in a poster session, and write a story about what life would be like if their feet were magnets.

Suggested Grade Levels: K–2

Lesson Objectives Connecting to the Standards

Content Standard A: Scientific Inquiry

- Use data to construct a reasonable explanation.
- Communicate investigations and explanations.

Content Standard B: Physical Science

- Understand that magnets attract and repel each other and certain kinds of other materials.
- Understand that objects can be described by the properties of the materials from which they are made and those properties can be used to separate or sort a group of objects or materials.

Featured Picture Books

Title	*That Magnetic Dog*	*Magnetic and Nonmagnetic*
Author	Bruce Whatley	Angela Royston
Illustrator	Bruce Whatley	Maria Joannou and Sally Smith
Publisher	Angus & Robertson	Heinemann Library
Year	1994	2003
Genre	Story	Non-narrative Information
Summary	Skitty is a dog with "magnetic" qualities. She doesn't attract metal, like keys and spoons. She attracts food.	This book explains what magnets do, magnetic and non-magnetic materials, and uses of magnets.

Time Needed

This lesson will take several class periods. Suggested scheduling is as follows:

Day 1: **Engage** with *That Magnetic Dog* read aloud, and **Explore** with What Can You Catch With a Magnet.

Day 2: **Explain** with Our Hypothesis and *Magnetic and Nonmagnetic* read aloud.

Day 3: **Elaborate** by Revisiting *That Magnetic Dog,* and **Evaluate** with The Day My Feet Were Magnets.

Materials

Magnet Warning Signs page (following lesson)

Per team of 3 or 4 students:

> 1 magnetic wand
>
> 1 pencil
>
> 20 cm piece of string
>
> Shoeboxes containing items to test for magnetic properties (suggested items: aluminum foil, paperclips, metallic-looking fabric, cotton balls, pennies, plastic and metal spoons, toothpicks, iron nails, steel wool)
>
> 1 piece of poster board
>
> Tape
>
> Crayons or markers
>
> Drawing paper
>
> Assortment of keys

Magnetic wands are available from

Sheridan WorldWise

1-800-433-6259

www.classroomgoodies.com

Student Pages

What Can You Catch With a Magnet?

Our Hypothesis

Background

The National Science Education Standards suggest that students in grades K–4 participate in explorations in which they learn that magnets attract other magnets and certain kinds of materials without touching them. Magnetism is a force acting at a distance. Although physicists consider "magnetic" materials to be those that can be magnetized, we are simplifying this classification for young learners. In the nonfiction book *Magnetic and Nonmagnetic,* any material that is attracted to a magnet is referred to as a *magnetic* material. Any material that is not attracted to a magnet is referred to as a *nonmagnetic* material. A common misconception is that all metals are magnetic. In fact, only materials that contain iron, cobalt, or nickel are attracted to a magnet. Steel is mostly iron, so steel is also magnetic. This lesson focuses on iron and steel as the most common magnetic materials because items made of cobalt and nickel are not nearly as common or familiar to students.

Magnets are used in a wide variety of electronic equipment. Placing a magnet close to such equipment may cause damage. Before using magnets in your classroom, make students aware that magnets

should be kept away from electronic equipment. Also, keep magnets away from credit cards, videotapes, and any other materials that have information on a magnetic strip.

engage

That Magnetic Dog Read Aloud

📖 Inferring

Show the book *That Magnetic Dog* to the class. Introduce the author and illustrator. Ask

? What do you think this book might be about?

Next, read the book aloud.

📖 Inferring

After reading, ask

? Why does the author describe the dog as "magnetic"?

? Have you ever heard the word *magnetic* used to describe someone?

? What does *magnetic* mean?

explore

What Can You Catch With a Magnet?

Before you begin this phase of the lesson, copy the magnet warning signs on fluorescent paper and post them on computers and other electronic equipment in the classroom. Explain to students that magnets can damage electronic equipment like televisions and computers, so they must pay attention to the signs. Then ask

"Fishing" with a magnet

? What types of items do you think are attracted to a magnet?

Put a variety of magnetic and nonmagnetic items (see materials list) in a shoebox for each team of students. Tell students that they are going to go fishing with a fishing pole made out of a magnet to see which items in the box they can catch. Show students how to make the fishing pole by tying one end of a string around the end of a magnetic wand and the other end of the string to a pencil.

Before they begin fishing, have students draw the items in the shoebox on the What Can I Catch With a Magnet? student page and then predict which items will be attracted to a magnet by circling them. After they have made their predictions, ask

? Why do you think those items will be attracted to the magnet?

Students are now ready to use their fishing poles to see what items in the box their magnets will catch. Tell students to keep the items caught by the magnet outside the box and leave the other items inside the box.

explain
Our Hypothesis and *Magnetic and Nonmagnetic* Read Aloud

In teams, have students create a poster to present their findings to the rest of the class. They can do so by drawing a line down the center of the poster board and taping the items that were caught by the magnet on one side of the poster board. They can tape the items that were still in the box on the

Making a poster

other side of the poster board. After they create their posters, ask

? What items did the magnet pick up?

? What do those items have in common?

? Were you surprised by any of the results?

Pass out the Our Hypothesis student page. Have each team develop a one-sentence hypothesis about what the items that were caught by the magnet have in common. They can then present the hypothesis, along with their poster, to the rest of the class. At this point, many students will propose that the items that are attracted to the magnet are metal and the items that are not attracted are not metal. Although this conclusion is not entirely correct, they are getting closer to the secret of magnetic materials. Show them that a penny is made of metal, but it is not attracted to the magnet. Some metals are attracted to a magnet, but not all. Some students may hypothesize that there is something special inside the items that were attracted to the magnet.

Using Features of Nonfiction/ Chunking

Show students the cover of the book *Magnetic and Nonmagnetic,* and then show the table of contents and a few of the inside pages. Ask

? Is this a fiction or nonfiction book? (nonfiction)

? How can you tell? (Answers may include: It has a table of contents, photographs, bold-print words, and diagrams.)

Explain that, because the book is nonfiction, you can enter the text at any point. You don't have to read the book from cover to cover if you are looking for specific information. Tell students that this book might be able to help them discover the secret of the materials the magnet picked up. Ask students to signal (by thumbs up, touching the nose, or some other method) when they hear the answer to the question.

? What special material is in the items that the magnet picked up?

Then read aloud pages 4–11 in *Magnetic and*

Nonmagnetic. Students should now understand that the items were picked up by the magnet *because they contain iron or steel.* That's the secret! Ask

? What do we call items that are attracted to a magnet? (magnetic)

? What do we call items that are not attracted to a magnet? (nonmagnetic)

Rereading

Tell students that, often, good readers will go back to a book and reread a section to be sure that they understand the information they read. Reread pages 6–11 and have students listen for the answers to those two questions. After listening, students can now label the side of the poster containing items that were attracted to a magnet with the word *magnetic,* and the other side with the label *nonmagnetic.*

elaborate

Revisiting *That Magnetic Dog*

Synthesizing

Reread the first page of *That Magnetic Dog* which states, "Magnets attract metal objects like keys and spoons." Based on what students have learned through the exploration phase of this lesson, they should be able to identify and correct the inaccurate information in that sentence. Ask

? Is it entirely correct to say, "Magnets attract metal objects?" (no)

? How could we rewrite that sentence to make it more scientific? (Magnets attract metal objects that contain iron or steel.)

Students should understand that magnets do not attract all metal objects. Demonstrate with a magnetic wand that most keys are not attracted to a magnet because most keys are not made of iron or steel. You may be able to find some old keys that do contain iron, but most are now made with zinc to prevent rusting.

Writing a class story

evaluate

The Day My Feet Were Magnets

Ask students to think about what would happen if they woke up one morning to find that their feet had mysteriously been turned into magnets. Write a story about it together on chart paper or on the board. (You may want to have older students write their own story.)

Ask

? What things would be attracted to your feet? Why?

? What things would not be attracted to your feet? Why?

? What kinds of problems would you have if your feet were magnets?

? What kinds of advantages or special powers might you have?

📖 Visualizing: Sketch to Stretch

Have each student illustrate the class story. Reread the story several times, and tell students that their drawings should show the sequence of events in the class story.

Inquiry Place

Have students brainstorm testable questions such as:

? Can magnetism pass through water?

? Can magnetism pass through wood, plastic, or other materials?

? Are larger magnets stronger than smaller magnets?

? What happens when two magnets touch?

Then have students select a question to investigate as a class, or have groups of students vote on the question they want to investigate as a team. After they make their predictions, have them design an experiment to test their predictions. Students can present their findings in a poster session or gallery walk.

More Books to Read

Bryant-Mole, K. 1998. *Magnets.* Chicago: Heinemann Library.

Summary: Text and experiments introduce the scientific properties of magnets, examining such topics as their strength, magnetic poles, and the making of magnets.

Branley, F.M. 1996. *What makes a magnet?* New York: HarperCollins.

Summary: This Let's-Read-and-Find-Out Science book explains how magnets work and includes instructions for making a compass.

Rosinsky, N. 2003. *Magnets: Pulling together, pushing apart.* Minneapolis: Picture Window Books.

Summary: Simple text and illustrations, accompanied by fun facts, explain how magnets work, why Earth is really a giant magnet, how a compass works, and more. Includes simple experiments, table of contents and glossary, and a website with links to other safe, fun websites related to the book's content.

NO MAGNET ZONE

KEEP MAGNETS AWAY

Name: _____

What Can You Catch With a Magnet?

Directions: Draw the items in the box. Circle the items you predict the magnet will catch.

Names: _____

Our Hypothesis

We think the items were attracted to the magnet because...

_____

_____

_____

Roller Coasters

Description

Learners explore ways to change the speed and direction of a rolling object by building roller coasters out of pipe insulation. They investigate the idea that gravity affects all objects equally by conducting dropping races with everyday items.

Suggested Grade Levels: K–4

Lesson Objectives Connecting to the Standards

Content Standard A: Scientific Inquiry

- Ask a question about objects, organisms, and events in the environment.
- Plan and conduct a simple investigation.

Content Standard B: Physical Science

- Understand that the position and motion of an object can be changed by pushing or pulling.

Featured Picture Books

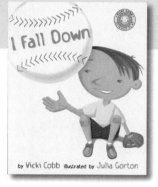

Title	*Roller Coaster*	*I Fall Down*
Author	Marla Frazee	Vicki Cobb
Illustrator	Marla Frazee	Julia Gorton
Publisher	Harcourt	HarperCollins
Year	2003	2004
Genre	Story	Non-narrative Information
Summary	Twelve people set aside their fears and ride a roller coaster, including one who has never done so before.	Simple experiments introduce the basic concept of gravity and its relationship to weight.

Time Needed

This lesson will take several class periods. Suggested scheduling is as follows:

Day 1: **Engage** with *Roller Coaster* read aloud. **Explore/Explain** with Roller Coaster Design Challenges.

Day 2: **Elaborate** with *I Fall Down* read aloud and Dropping Races.

Day 3: **Evaluate** with Falling Objects Quiz.

Materials

Roller Coaster supplies for each pair of students:

> 6 ft. length of foam insulation to fit a 1 in. pipe (split lengthwise) with a plastic 20 oz. cup taped to the end

> Ball that will roll in the split pipe insulation prepared above, such as a foosball, large marble, or ball bearing.

Supplies to use during *I Fall Down* read aloud:

> Penny

> Key

Dropping races supplies for pairs of students: tennis ball, marble, paperclip, penny, book

> Notebook paper

> Dry sponge

> Bar of soap

> Heavy shoe

> Lightweight shoe

> 2 identical large rubber bands

Student Pages

Roller Coaster Challenges

My Roller Coaster

Dropping Races

Falling Objects Quiz

Background

The National Science Education Standards recommend that students explore and describe motion by pushing, pulling, throwing, dropping, and rolling a variety of everyday objects. The Standards suggest that K–4 students begin to focus on the position and motion of objects as well as the motion and forces required to control the objects. By making careful observations and recording data, students in even the earliest grades can begin to look for patterns in their work with motion and can determine the speed of an object as fast, faster, or fastest. In this lesson, students investigate how to control the speed of a model roller coaster and discover how gravity affects the motion of objects as they fall.

Gravity is a force that pulls all objects toward the center of the Earth. Earth's gravity keeps us on the ground and causes objects to fall. It also keeps the Moon in orbit. The Sun's gravity keeps the planets in orbit around it. Although most roller coasters are pulled up the first hill by a chain, what you may not realize as you're cruising down the track at 60 miles an hour is that the coaster has no engine. Gravity is the main force responsible for the movement of the roller coaster. Most of the time, the first hill on a roller coaster slopes down about 50 degrees. This is the most exciting drop of the ride! The roller coaster goes faster and faster the closer it gets to the ground.

One common misconception that many children and adults have about Earth's gravity is that heavier objects fall faster than lighter objects. This is not true. Gravity affects all objects equally no matter how much they weigh. We all know that a feather falls slower than a hammer when dropped on Earth. This is because the feather is more affected by *air resistance.* If you could get rid of the air, the hammer and feather would hit the ground at the same time. The astronauts on the Apollo 15 Mission proved this to be true by dropping a feather and a hammer on the Moon from the same height at the same time. Both hit the ground at the same time. You can see actual video footage of the Apollo 15 astronauts dropping a feather and a hammer on the Moon at: *http://nssdc.gsfc.nasa.gov/planetary/lunar/ apollo_15_feather_drop.html.*

engage
Roller Coaster Read Aloud

📖 Making Connections: Text to Self/Turn and Talk

Show students the cover of the book, *Roller Coaster.* Introduce the author and illustrator, Marla Frazee. Tell students that Marla Frazee has three sons who love roller coasters. One summer, their family spent a week on a driving vacation and the whole time her boys talked about roller coasters: which boy was bravest, which coaster was scariest, which drop was highest. This gave Marla Frazee the idea for making this book. (There is more information about Marla Frazee and her family on the book jacket and at *www. marlafrazee.com.)*

Before reading, ask

? Have you ever been on a roller coaster? What was it like? If you've never been on one, what do you think it would be like?

Have students turn and talk to a partner.

Roller coaster *read aloud*

📖 Inferring

Begin reading the book, but stop after reading pages 14 and 15, where the roller coaster is slowly going up the hill. Ask

? What do you think the next picture in the book will look like?

Have students turn and talk to a partner.

Visualizing: Sketch to Stretch

Continue reading the book, but stop after reading page 27 ("Wheeeeeee!"). Have students close their eyes and imagine what it would feel like to be on the roller coaster in the book. Ask

? How would you feel if you were on this roller coaster?

? What do you think your face would look like if you were riding this roller coaster?

Have students make a sketch on a sticky note of what they think their face would look like if they were on the roller coaster. They can share their picture with a partner. (For fun, have the whole class make their roller coaster faces on the count of three.) Then, finish reading the book aloud.

explore/ explain

Roller Coaster Design Challenges

Announce to students that they are going to work with a partner to design their own roller coaster. Hold up a piece of foam pipe insulation with a plastic cup taped to one end. Tell them that this will be the track. Show students a ball and tell them that this will be the roller coaster car. Caution them not to throw the ball or push it down the track. They should simply release the ball and let it roll. Show them the cup and ask

? What do you think the cup is for? (to catch the ball or to stop the ball from rolling away)

Give each pair of students these supplies and the Roller Coaster Challenges student page. Have them complete the challenges below:

? Can you make the ball roll from one end of the track and stop in the cup?

? Can you make the ball roll faster?

? Can you make the ball roll more slowly?

? Can you make the ball go over a hill on your roller coaster?

? Can you make the ball go over two hills on your roller coaster?

? Can you make the ball go through a loop on your roller coaster?

Allow students several minutes to work on the challenges. Encourage them to make observations about where on the track the ball moves fastest and slowest. Then bring students back together and ask

? How did you make the ball roll faster? (by raising one end a lot higher than the other)

? How did you make the ball roll more slowly? (by raising one end only a little higher than the other)

? How did you make a hill on your roller coaster? (by bending the middle up) Were you able to make two hills?

? Which was the highest, the first hill or the second hill? (The first hill had to be the highest to get the ball going fast enough to go over the second hill.)

? How did you make the ball go over the hills or around loops on your roller coaster? (by making the beginning of the track steep)

? Did the ball ever fall off of the roller coaster? What made it fall?

? What causes the ball to go down the track? (Answers may vary; the next activity will introduce students to the concept of gravity.)

Next, hand out the My Roller Coaster student page. Tell students that they will draw a roller coaster using what they have learned from making their model roller coasters. Tell them they can make as many hills and loops on their roller coaster track as they wish, as long as they think the roller coaster would actually work in real life. Encourage them to use color and add details to their roller coaster car. They can even draw people in it if they like. Tell students to label where the roller coaster would be moving the fastest and where it would be moving the slowest.

Before they begin, ask

Making "roller coaster faces"

? What do you need at the beginning of the ride to get the roller coaster car moving fast? (a high hill)

? Can a second hill be higher than a first hill? (No, the first hill has to be the highest.)

When they are finished, evaluate their understandings about motion by asking them questions such as

? Where on the track does your roller coaster car move the fastest? (toward the bottom of the hills)

? Where on the track does your roller coaster car move the slowest? (toward the top of the hills)

elaborate

I Fall Down Read Aloud

Explain to students that there is a force that pulls everything toward the ground. On Earth, it is impossible to escape the pull of this force. It affects everything we do every day of our lives. In the case of our model roller coaster, this force pulls the ball toward the ground. Tell students that you have a book that will tell them more about this incredible force!

Note: Vicki Cobb suggests that the best way to use her book *I Fall Down* is to do the activities described in the book, without rushing, as they

Making a loop

My roller coaster

experience. See "Note to the Reader" on page 3 of *I Fall Down*.

Inferring

Introduce the author and illustrator of *I Fall Down*. Ask

? Look at the title and the picture on the cover. What do you think this book might be about?

Determining Importance

Ask students to signal (by touching an ear, raising a hand, or some other method) when they hear the name of the force that makes things fall. Then read pages 1–11 (stop reading after "down, down, down") and ask

? What's the name of the force that is always pulling things "down, down, down"? (gravity)

Continue reading aloud to page 15 (stop reading after the pages that describe dropping the penny and the key). Hold a penny and a key in the same hand and ask

come up during the reading. Before you begin reading, make sure you have all the necessary supplies at hand. The author also suggests not turning the page to the explanation until *after* the child has made the discovery. That way, the book will reinforce what the child has discovered through

Dropping races

? What will happen if I open my hand? (The penny and key will fall.)

? What causes them to fall? (Gravity pulls them down.)

? Which one will hit the ground first? (Answers will vary.)

Ask students to watch the ground closely as you open your hand to see which object hits the ground first. Be sure that the penny and key are released at the same exact time. Students should notice that the penny and key hit the ground at the same time. You may need to do this several times to convince students!

Dropping Races

Pass out the supplies for dropping races (tennis balls, marbles, paper clips, pennies, books) and the Dropping Races student page, and allow students time to try all of the dropping races.

Have students share their results for the dropping races. Students should discover that all of the races result in a tie. If students are not convinced

that all races resulted in a tie, perform some whole-class demonstrations of the dropping races.

Then read pages 16 and 17 aloud (about dropping races). Explain that gravity affects all objects equally. This means that heavy objects fall at the same rate as lighter objects. Some students may ask about objects like a feather or a piece of paper that they have observed falling slowly. Tell students that when you drop something that the wind could easily blow away, it does fall slowly because, as mentioned on page 17, air is "fighting" against gravity with very light objects. Drop a sheet of notebook paper and have students observe the paper "fighting" against gravity as it slowly drifts down.

Continue reading through page 19 "Astronauts proved this on the Moon … every dropping race was a tie." Some students may have trouble believing that, if there were no air, every dropping race would be a tie. One way to demonstrate this is to do the following additional activity (not described in the book):

1 Take a book and a piece of paper smaller than the book, and drop them at the same time from the same height. Students will observe that the paper falls more slowly. Explain that because the paper is lighter and more spread out, air "fights" against gravity.

2 Ask

? What do you think would happen if I put the paper on top of the book and dropped them together so that the book pushes the air out of the way?

Have students make their predictions.

3 Place the paper on top of the book and drop them together. The book and paper land together, because the book is "fighting" the air, not the paper.

Read pages 20 and 21 ("Which hits your hand harder, the sponge or the soap?"), and then demonstrate the activity with a dry sponge and a bar of soap. Ask for a student volunteer to hold his or her hand outstretched and palm up. Drop

A dropping race demonstration

the sponge into the student's hand. Repeat with the bar of soap. Ask

? Which hits harder, the sponge or the soap? (the soap)

? Why? (because it is heavier)

Next, read pages 22–29 aloud. Show students an example of a heavy shoe and a light shoe tied to rubber bands. Ask students

? Which shoe is heavier? (The one that stretches the rubber band the longest.)

Read the rest of the book aloud. Ask

? How much would you weigh if there was no gravity? (Nothing, without gravity we would all be weightless.)

evaluate

Falling Objects Quiz

Review the concepts that have been explored in this lesson, and then give students the Falling Objects Quiz. Answers are:

1. c
2. b
3. b
4. b
5. a
6. b

Inquiry Place

Have students brainstorm testable questions such as

? Which rolls faster, a heavier ball or a lighter ball?

? Which surface lets a ball roll farthest?

? Which falls faster, a feather or a sheet of paper?

? How can you slow the fall of an object?

Then have students select a question to investigate as a class, or have groups of students vote on the question they want to investigate as a team. After they make their predictions, have them design an experiment to test their predictions. Students can present their findings at a poster session or gallery walk.

More Books to Read

Berenstain, J. and S. 1998. *The Berenstain Bears ride the Thunderbolt*. New York: Random House Books for Young Readers.

Summary: Readers will love spending a day at the Bear Country Amusement Park, where they'll experience the stomach-dropping, heart-stopping thrills of a giant roller coaster right along with the Berenstain Bears.

Cole, J. 1998. *Magic school bus plays ball: A book about forces*. New York: Scholastic.

Summary: Mrs. Frizzle and her class shrink to fit inside a physics book where they enter a page about a baseball field with no friction. The kids learn about how throwing, running, and catching would work in a world without friction.

Llewellyn, C. 2004. *And everyone shouted "Pull!": A first look at forces and motion*. Minneapolis: Picture Window Books.

Summary: Hop on the cart, and join the farm animals as they find out how to take their heavy load on the hilly journey to market.

Stille, D. 2004. *Motion: Push and pull, fast and slow*. Minneapolis: Picture Window Books.

Summary: Up, down, forward, and back. Sideways or around and around. See how things get moving—and what makes them stop—in this lively book on motion. Includes a table of contents, glossary, simple experiments, and a FactHound website with links to other safe, fun websites.

Whitehouse, P. 2003. *Rolling*. Chicago: Heinemann Library.

Summary: Brief text, colorful photographs, and simple, hands-on activities explain the properties that make rolling easy or difficult.

Websites

Funderstanding Roller Coaster
www.funderstanding.com/k12/coaster

Amusement Park Physics: What are the forces behind the fun?
www.learner.org/resources/series136.html

Name: _____

Roller Coaster Challenges

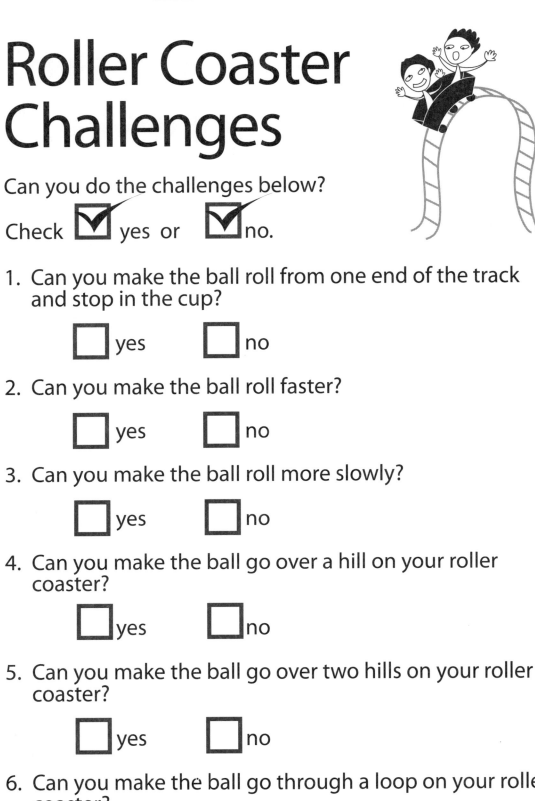

Can you do the challenges below?

Check ☑ yes or ☑ no.

1. Can you make the ball roll from one end of the track and stop in the cup?

    ☐ yes   ☐ no

2. Can you make the ball roll faster?

    ☐ yes   ☐ no

3. Can you make the ball roll more slowly?

    ☐ yes   ☐ no

4. Can you make the ball go over a hill on your roller coaster?

    ☐ yes   ☐ no

5. Can you make the ball go over two hills on your roller coaster?

    ☐ yes   ☐ no

6. Can you make the ball go through a loop on your roller coaster?

    ☐ yes   ☐ no

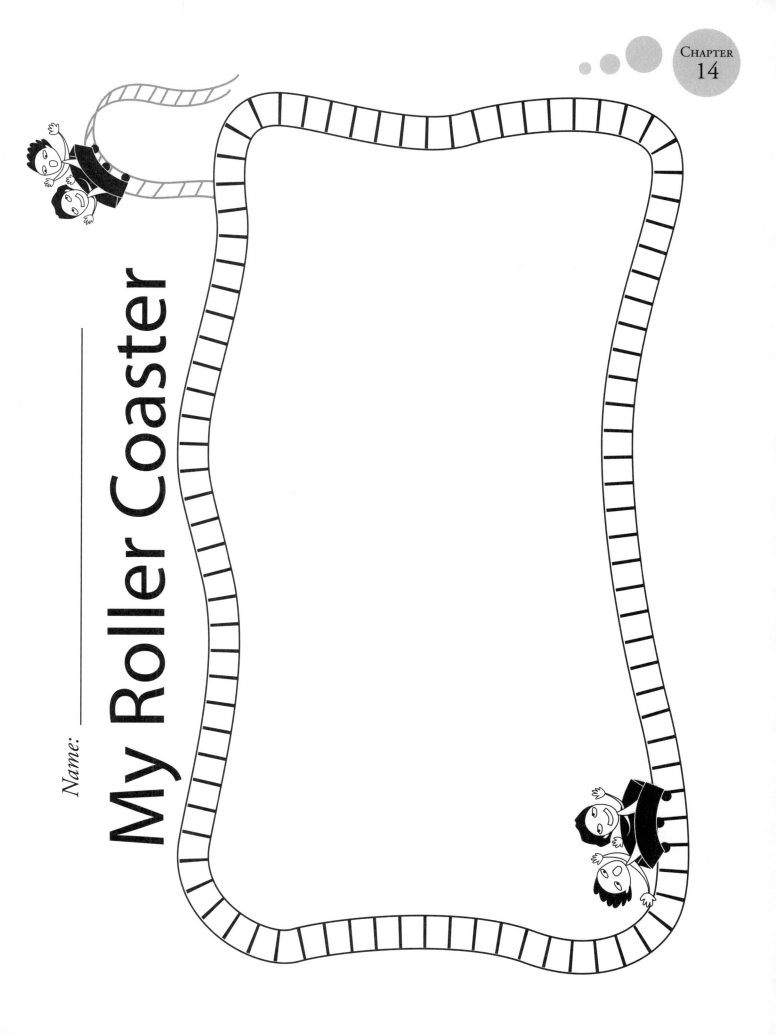

Name: _____

My Roller Coaster

Name: _____

Dropping Races

Drop the following objects at the same time from the same height. Check ☑ the winner of each dropping race.

1. Tennis Ball ☐   Marble ☐   Tie ☐

2. Tennis Ball ☐   Paper clip ☐   Tie ☐

3. Tennis Ball ☐   Penny ☐   Tie ☐

4. Penny   ☐   Book ☐   Tie ☐

5. Paper clip ☐   Book ☐   Tie ☐

Falling Objects Quiz

Anna drops a bowling ball and a paper clip from the same height at the same time.

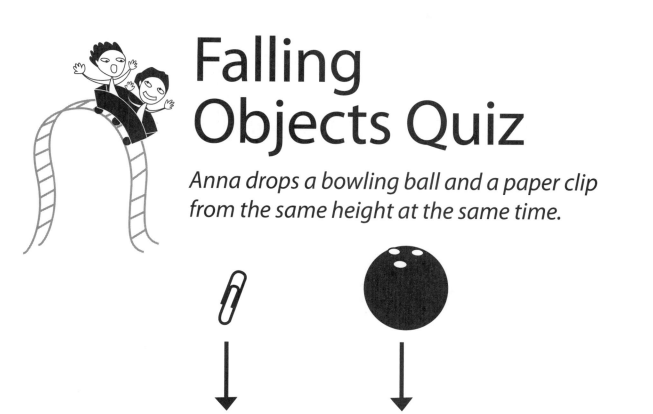

ground

1. Which will hit the ground first?
   a. paper clip
   b. bowling ball
   c. Both will hit at the same time.

2. Which will hit the ground the **hardest**?
   a. paper clip
   b. bowling ball
   c. Both will hit just as hard.

3. What force causes objects to fall?
   a. friction
   b. gravity
   c. air

Falling Objects Quiz *cont.*

Jesse tied the same rubber band to two of his toys.

4. Which toy is the heavier?
   a. the toy elephant
   b. the toy truck
   c. They weigh the same.

5. Gravity is always:
   a. pulling things
   b. pushing things
   c. lifting things

6. Where on the track below would a roller coaster be going the fastest? Circle the letter.

Start

Mirror, Mirror

Description

Using flashlights, mirrors, and spoons, learners explore how light travels. They investigate how light is reflected differently by curved and rough surfaces and why mirrors are the best surfaces for seeing themselves.

Suggested Grade Levels: K–2

Lesson Objectives Connecting to the Standards

Content Standard A: Scientific Inquiry

- Ask a question about objects, organisms, and events in the environment.
- Design and conduct simple experiments to answer questions.
- Use data to construct reasonable explanations.

Content Standard B: Physical Science

- Understand that light travels in a straight line until it strikes an object. Light can be reflected by a mirror, refracted by a lens, or absorbed by the object.

Featured Picture Books

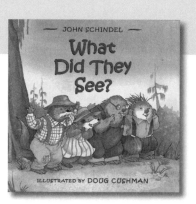

Title	*What Did They See?*	*I See Myself*
Author	John Schindel	Vicki Cobb
Illustrator	Doug Cushman	Julia Gorton
Publisher	Henry Holt	HarperCollins
Year	2003	2002
Genre	Story	Non-narrative Information
Summary	Raccoon hurries to show Beaver, Porcupine, and Otter the most amazing "thingamajig" that they have ever seen (a mirror).	Learn why you can see yourself in shiny objects with this fun, interactive book about light.

Time Needed

This lesson will take several class periods. Suggested scheduling is as follows:

Day 1: **Engage** with *What Did They See?* read aloud, and **Explore/Explain** with Can You See Yourself?

Day 2: **Explore/Explain** with Mirror Challenges and *I See Myself* read aloud.

Day 3: **Elaborate/Evaluate** with Silly Spoons.

Materials

Per pair:

Plastic mirror hidden in a "mystery envelope" with a question mark on the envelope

Flashlight

Shiny metal spoon

Crinkled aluminum foil square (about 5 cm by 5 cm)

Ordering Information for Mirrors

Sheridan WorldWise

www.classroomgoodies.com

1-800-433-6259

Package of 15 small mirrors

Order # WNB 7641

Student Pages

Mirror Challenges

Silly Spoons

Background

Understanding that light travels in a straight line until it strikes an object and that light can be reflected by a mirror, refracted by a lens, or absorbed by an object are key components to the Physical Science Standard in grades K–4. In this lesson, students observe the path of light as it reflects off a mirror and what happens when light hits other objects.

Light is an essential part of our everyday lives. Without light, we would not be able to see. This fact can be difficult for some students to believe, because most of us have never been in a completely dark place before. Light behaves according to special rules. For example, it always travels in a straight line until it hits something. When light hits a mirror, it *reflects*, or bounces off the surface. (Note: The book *I See Myself* uses the kid-friendly term *bounce* instead of the scientific term *reflect*. Technically speaking, light does not "bounce" in the same way that a ball bounces. Instead, light is actually absorbed by the molecules in the mirror and then sent back out as a reflection.)

The *law of reflection* states that, when something bounces off a perfectly flat surface, the angle at which it hits the surface will be equal to the angle at which it bounces away. A *mirror* reflects light in this way because of its very flat, smooth surface. A mirror is made of a glass sheet in front of a metallic coating where the reflection actually occurs. A *curved mirror,* like a funhouse mirror, can be thought of as consisting of a very large number of small flat mirrors oriented at slightly different angles. The

law of reflection still applies, but the image you see is distorted. A very shiny spoon can be used to demonstrate this type of distortion.

Light isn't just reflected off mirrors; light is reflected off every object you see. When light strikes a rough surface, it reflects off in many directions due to the microscopic irregularities of the surface. Thus, a mirror image is not formed. This is called *diffuse reflection*. As you are reading this, light is reflecting off of the page. The black type on the page is absorbing all the light that hits it, but the rest of the page is reflecting light. The reflected light is scattering in many directions. Some of the scattered light reaches your eyes. That's why you can read this, but you can't see your reflection in the page.

engage
What Did They See? Read Aloud

Note: If you are unable to locate a copy of *What Did They See?*, you can use a riddle to replace the read aloud: "It is flat. It is shiny. It can follow your every move. It contains something lovely, good looking, and amazing. What is it?" Then have them open their envelope to see the mirror.

Inferring

Hold up the cover of the book *What Did They See?* Ask

? What do you think this book might be about? Why do you think so?

? Do you think the book is fiction or nonfiction? How can you tell? (Possible answers include: It has cartoon animals on the cover, and the animals are wearing clothing.)

Questioning

Read *What Did They See?* to the class. Model the thinking of a good reader by saying aloud "I wonder what it could be?" when appropriate in the story. Stop reading before you read the last page, and pass out a sealed mystery envelope with a mirror in it to each pair of students. Then read the last page and tell them that the "good-looking thingamajig, lovely whatchamacallit, dashing

thingamabob" is in the envelope. Have them take it out as you read, "and what each one saw was something wonderful indeed." Ask

? What wonderful thing do you see when you look at the amazing "thingamabob" in the mystery envelope? (myself)

? Why did each character in the story think that what they saw was so "dashing," "amazing,"

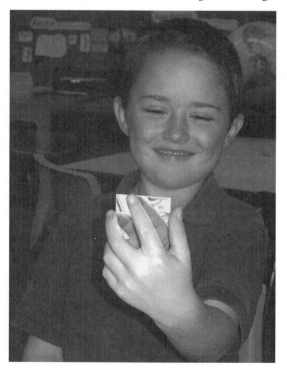

Looking at the "amazing thingamabob" from the mystery envelope

"good looking," or "lovely"? (They were seeing themselves in the mirror.)

? How do you know this book is fiction? (Possible answers include: Animals can't really talk, and they don't wear clothes.)

explore/explain
Can You See Yourself?

Next, give students a few minutes to walk around the room to discover if they can see themselves in anything. Ask

? In what things could you see yourself? (Answers include: metal objects, doorknobs, the glass in a picture frame, and windows.)

? What do these things have in common? (Answers include: They are smooth, and they are shiny.)

? Did you appear as clear in these objects as you appeared in the mirror? (no)

? Why do you think the mirror was best for seeing yourself? (Answers will vary.)

explore/explain
Mirror Challenges

Have each pair of students use their mirror and a flashlight for this activity. Make your room as dark as possible. Have students shine the flashlight on the mirror and observe where the beam of light goes. Ask

? Does the light from the flashlight go through the mirror? (No. It bounces off.)

? Where does the light go after it bounces off the mirror? (Answers might include: behind me and on the wall.)

Keep the room darkened, and pass out the Mirror Challenges student page. Have students put a checkmark in each box as they complete challenges 1–4. (They can use their flashlights to

view their papers.) As they are working, circulate to ask questions such as

? What do you think?

? How do you know?

? What is your evidence?

Turn the lights back on, and have students complete challenges 5–7.

explain
Mirror Challenges Discussion and *I See Myself* Read Aloud

After students have completed the Mirror Challenges student page, ask them

? Were you able to do all of the challenges?

? Which one was the most difficult? Why?

? Did any of the results surprise you? Why?

? What word appeared in the mirror? ("REFLECT")

? How did your name appear in the mirror? (backwards, or backwards and upside down, depending on where they placed the mirror.)

Tell students that the word *reflect* means to send back. Ask

? When you look into a mirror, what reflects or is sent back? (light)

Then explain that, when we look into a mirror, we call our mirror image a *reflection,* because it is formed by light reflecting off the mirror to our eyes.

Ask

? Do you think you could see your reflection in the mirror if this room were completely dark? (Answers will vary.)

Explain that, if the room were dark, there would be no light to reflect. In fact, you cannot see anything without light. Challenge students to

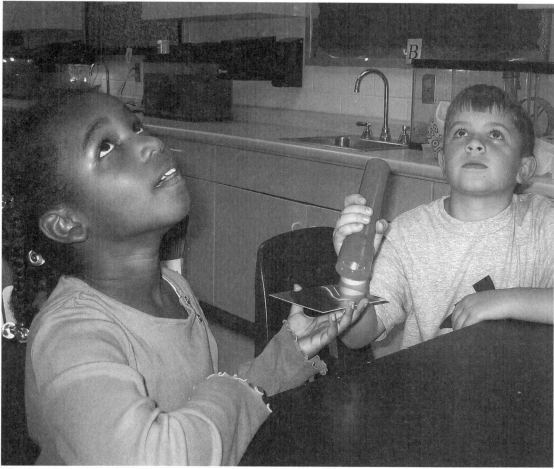

Mirror challenges

try this at home with adult supervision:

Take a flashlight and a small hand mirror into a room with no windows, such as a closet or bathroom. Shut the door, and cover the cracks below the door with towels. Then turn off the light and the flashlight. Look into the mirror. What can you see?

Determining Importance

Tell students you have a nonfiction book to help them learn about how mirrors work. Have them listen to find out why mirrors seem to be the best objects for seeing themselves. Read *I See Myself* to the class.

After reading, ask

? Why are mirrors the best objects for seeing yourself? (because they are flat and shiny)

? In order to see yourself, in order to see anything, you must have what? (light)

? When a ray of light hits a mirror, what happens? (It makes a perfect "bounce" every time.)

? Why can't you see yourself in a sweater or in the pages of a book? (A sweater or page is not perfectly flat and shiny. The light reflects off the sweater or page and scatters in many directions.)

Silly spoons

elaborate/evaluate
Silly Spoons

Ask students

? Have you ever looked into a fun-house mirror or into a mirror that made your reflection look funny? What did that mirror look like? (It was curved or not completely flat.)

Give each student the Silly Spoons student page. Give each pair of students a shiny metal spoon, a small plastic mirror, and a square of crinkled foil. Have them complete the activities on the student page and then share their explanations with other pairs. Discuss their findings as a class.

Answers to the Silly Spoons student page are as follows:

1 and **2** Answers might be: My face looks funny, and my nose looks too big, OR my face looks upside down (answer depends on which way the student is holding the spoon).

3 It is not flat.

4 yes

5 no

6 b. (Explain that because of the crinkles in the shiny foil, the light rays are reflected in many directions. Even though the foil is shiny, you don't see a clear reflection because it is not smooth.)

7 light, smooth surface, shiny surface

Inquiry Place

Have students brainstorm testable questions about light, such as

? What material lets more light through: tissue paper or construction paper?

? What time of day is your shadow the longest? the shortest?

? Can you see your shadow in the shade of a tree?

Then have students select a question to investigate as a class, or have groups of students vote on the question they want to investigate as a team. After they make their hypotheses, have them design an experiment to test their hypotheses. Students can present their findings at a poster session.

More Books to Read

Bahrampour, A. 2003. *Otto: The story of a mirror*. New York: Farrar, Straus and Giroux.

Summary: A humorous story about Otto, a mirror in a hat shop, who dreams of faraway lands. He eventually escapes from the shop and ends up on an exotic island where he meets another mirror and they reflect back and forth forever and ever.

Bulla, C.R. 1994. *What makes a shadow?* New York: HarperCollins.

Summary: This Let's-Read-and-Find-Out Science book gives simple explanations for shadows. Each page offers a brief description of an object and its shadow. Simple activities to demonstrate various sizes and shapes of shadows are included.

Narahashi, K. 1987. *I have a friend*. New York: Aladdin.

Summary: A small boy tells about his friend who lives with him, who follows him, who sometimes is very tall, but who disappears when the Sun goes down—his shadow.

Pulley Sayre, A. 2002. *Shadows*. New York: Henry Holt.

Summary: A boy and girl search for shadows on a sunny summer day.

Rosinsky, N.M. 2003. *Light: Shadows, mirrors, and rainbows.* Minneapolis: Picture Window Books.

Summary: Simple text and fun, colorful illustrations help readers understand how shadows are made, how mirrors work, how rainbows are made, and more. Includes simple experiments, table of contents and glossary, and a website with links to other safe, fun websites related to the book's content.

Name: _____

Mirror Challenges

Put a check mark in the box ☑ *after*
you complete each challenge.

☐ 1. Using the mirror, make your light shine on the wall behind you.

☐ 2. Using the mirror, make your light shine on the ceiling.

☐ 3. Using the mirror, make your light shine on your shirt.

☐ 4. Join another team, and make your light bounce off of two mirrors at the same time.

☐ 5. Move your mirror around until you can read the word in the box below.

ЯEFLECT

Then write the word on this line: _____

 6. Write your name in the box below.

```




```

☐ 7. Show how your name appears in the mirror in the box below.

```




```

Silly Spoons

1. Look in the spoon. How does your face appear in the spoon?

   _____
   _____

2. Look in the other side of the spoon. How does your face appear in the other side of the spoon?

   _____
   _____

3. Why do you think the spoon does not give you a perfect reflection of yourself?

   _____
   _____
   _____

4. Look in the mirror and smile. Can you see your smile in the mirror? _____

   _____

5. Look in the crinkled foil and smile. Can you see your smile in the foil? _____

   _____

Silly Spoons cont.

6. Circle the picture that shows light bouncing off a crinkled piece of foil.

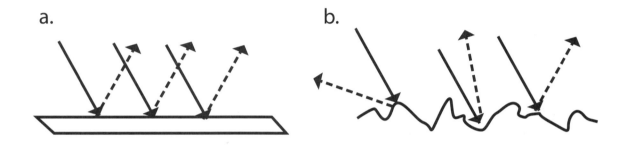

a.

b.

7. What three things are needed for you to see your reflection? (circle three things)

      light   dark

      colored surface   smooth surface

      shiny surface   bumpy surface

If You Find a Rock

Description

Learners observe, describe, and sort a variety of rocks, discovering that rocks have certain physical properties by which they can be classified.

Suggested Grade Levels: 2–4

Lesson Objectives Connecting to the Standards

Content Standard A: Scientific Inquiry

- Ask a question about objects, organisms, and events in the environment.
- Employ simple equipment and tools to gather data and extend the senses.
- Communicate investigations and explanations.

Content Standard D: Earth and Space Science

- Understand that Earth materials include solid rocks and soils and that they have different physical and chemical properties which make them useful in different ways.

Featured Picture Books

Title	*If You Find a Rock*	*Rocks: Hard, Soft, Smooth, and Rough*
Author	Peggy Christian	Natalie M. Rosinky
Illustrator	Barbara Hirsch Lember	Matthew John
Publisher	Harcourt	Picture Window Books
Year	2000	2003
Genre	Story	Non-narrative Information
Summary	Soft, hand-tinted photographs and simple, poetic text celebrate the variety of rocks that can be found, including skipping rocks, chalk rocks, and splashing rocks.	Simple text and cartoonish illustrations provide information on igneous, sedimentary, and metamorphic rocks.

Time Needed

This lesson will take several class periods. Suggested scheduling is as follows:

Day 1: **Engage** with *If You Find a Rock* read aloud and then have students bring in a special rock the next day.

Day 2: **Explore/Explain** with I Found a Rock and Rock Sorting.

Day 3: **Elaborate** with *Rocks: Hard, Soft, Smooth, and Rough* read aloud and rock identification.

Day 4: **Evaluate** with Pet Rock Posters.

Materials

Per student:

Hand lens

Centimeter ruler

A rock (brought in by student)

Colored pencils, crayons, or markers

Poster board

Per group of 5 students:

One of each of the following rock samples: obsidian, granite, sandstone, limestone, and marble

Per class:

1 interesting rock to engage students

Optional: Photo of the original Pet Rock packaging

Rock specimens in packs of 10 are available from *www.carolina.com*	
Rock	Order Number
Obsidian, Black	GEO1112B
Granite, Gray	GEO1080B
Sandstone, Red	GEO2012B
Limestone, Fossil	GEO1198B
Marble, White	GEO2054B

Student Pages

I Found a Rock

Pet Rock Advertising Poster Rubric

Background

Children are naturally curious about the world around them, including the rocks beneath their feet. Learning about the properties and uses of earth materials such as rocks helps young children build a foundation for later understandings about the interactions of the Earth's *geosphere* (crust, mantle, and core), *hydrosphere* (water), *atmosphere* (air), and *biosphere* (living things). The National Science Education Standards state that students in grades K–4 should understand that earth materials include solid rocks and soils, water, and the gases of the atmosphere. These materials have different physical and chemical properties that make them useful in different ways. The Standards also suggest that young children be encouraged to closely observe the objects and materials in their environment, note their properties, and distinguish them from one another. Following these suggestions, this lesson focuses primarily on recognizing properties of rocks (shape, size, color, texture, and luster, but not hardness,

which is a property used to identify minerals only), as well as understanding how properties of rocks can be used to sort them and exploring how a rock's properties and its uses are related.

In the Elaborate activity, students learn that rocks can melt inside the Earth, and that they are made of *minerals*. They are also introduced to the idea that rocks can be classified as igneous, sedimentary, or metamorphic depending upon how they are formed. *Igneous* rock was once melted within the Earth. After it forms, melted rock (*magma*) can push its way upward through cracks in the Earth's crust and cool while still within the Earth to form *plutonic* igneous rock or reach the surface where it is called *lava*. The lava then cools and hardens to become *volcanic* igneous rock. *Granite* is an example of plutonic igneous rock and *obsidian* is an example of volcanic igneous rock. Rocks can also form when *sediments* such as sand, mud, pebbles, bones, shells, and plants settle into layers on the bottoms of lakes, oceans, or rivers. Over millions of years, the top layers press down on the bottom layers and the bottom layers become *sedimentary* rock. *Sandstone* and *limestone* are examples of sedimentary rock. Limestone often contains the fossilized remains of animals that lived millions of years ago. The third kind of rock is *metamorphic* rock—rock that was formed when another kind of rock was exposed to tremendous heat and pressure over a long period of time. For example, *marble* is a metamorphic rock formed when limestone is "squeezed and cooked" inside the earth. The minerals within metamorphic rock are often arranged in stripes or swirls caused by heat and uneven pressure. Students learn that rocks can be identified by their color, texture, mineral arrangement, and luster, but not by their size or shape.

engage

If You Find a Rock
Read Aloud

With an interesting rock hidden in your hand, announce to the class that you have found something that is older than them, older than the school building, even older than you … something that could even be millions of years old! Have students guess what it is. Reveal the rock, and then tell students that a rock is probably one of the oldest things that they will ever touch. Ask students to share observations of the rock as it is passed around. Then tell students that you have a book to read to get them thinking about special rocks.

Making Connections:
Text to Self

Introduce the author and illustrator of the book, *If You Find a Rock*. The author, Peggy Christian, is a rock hound who was born in the Rocky Mountains

of Colorado and loves skiing, camping, and reading. Build connections to the author by asking

? What is a *rock hound*? (a person who likes to collect rocks)

? Is anyone here a rock hound?

? What do you call a scientist who studies rocks to learn about the Earth? (a geologist)

? Would you like to be a geologist?

Explain that there are many people, both men and women, who choose geology as a career and devote their entire lives to studying it. Tell students that Peggy Christian's father was a geologist and maybe that is why she loves rocks so much.

Determining Importance

Explain that, while you are reading the book aloud, you would like students to think about what some of the rocks in the book are used for and what properties, or characteristics, make them suited for those uses.

Read aloud *If You Find a Rock*. (For fun, stop after reading the page about the wishing rock

If You Find a Rock *read aloud*

and invite students to close their eyes and make a wish.)

After reading, ask

? What are some of the uses for the special rocks in the book? (Answers might include: wishing rock, skipping rock, chalk rock, resting rock, splashing rock, and worry rock.)

? Have you ever owned a special rock?

? What made it special to you?

Tell students that they are going to be rock hounds on the hunt for their own special rock. They can go outside and, with adult supervision, search for a rock or select a rock from their own collection. They should bring their special rock to school the next day. Send a letter home to inform parents of the assignment. Include these rules for students to follow: *Your rock must be smaller than a tennis ball. You are not allowed to throw your rock.* You may want to have extra rocks available for students who don't bring one in.

explore/explain

I Found a Rock

The next day, have students place their rocks on their tables or desks. Ask them to observe their own rock and then look around at some of the rocks near them. Discuss the following questions:

? How are the rocks alike?

? How are the rocks different?

Encourage students to notice that rocks come in a wide variety of colors, shapes, sizes, and other characteristics. Then explain that a scientific tool called a hand lens can help them get an even closer look at their rocks. Demonstrate the proper way to use a hand lens (holding the lens close to one eye while bringing the rock toward the hand lens until it comes into focus), and caution them that touching the rock to the surface of the hand lens can scratch the lens. Pass out hand lenses to all students, and have them use the lenses to observe their rock more closely.

Next, revisit the book, *If You Find a Rock*. Ask students to recall the rocks described in the book. List some of the rocks on the board, such as

- skipping rock
- chalk rock
- resting rock
- wishing rock
- worry rock
- climbing rock

Then ask

? What makes each rock in the book suited for its special use? (Answers might include: its shape, its color or size, and how it feels.)

Explain that these things—shape, color, size, texture (how it feels)—are called *properties* of rocks. Discuss how the properties of each rock in the book make it suited for a different purpose. For example, a skipping rock is used for skipping across water. The properties that make it suited

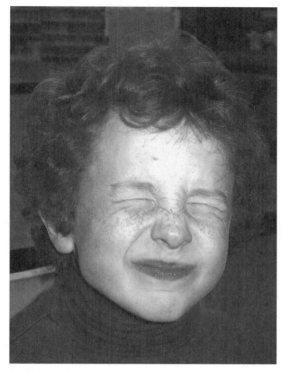

Making a wish

for that purpose are its flat and rounded *shape* and its small *size*. Shape and size are properties of rocks. A chalk rock is used to make pictures on the pavement. The properties that make it suited for that purpose are its white *color* and its soft, dusty *texture*. Texture describes how a rock feels. Color and texture are also properties of rocks. Another property students may notice as they observe their rocks is *luster*, or how the tiny specks in rocks reflect light. Words that describe luster include shiny, dull, and sparkly.

Explain that shape, size, color, texture, and luster are different properties of rocks that make each one unique. Geologists who study rocks use *some* of these properties to identify different types of rocks. Tell students that they are going to observe and record the properties of their own special rock. Pass out the I Found a Rock student page and centimeter rulers. Make sure students understand how to record the properties listed on the data table by discussing questions such as

? What are some words that might describe a rock's *color*? (Answers might include: black, white, and reddish-brown.)

? What are some words that might describe a rock's *texture*? (Answers might include: bumpy, smooth, and rough.)

? What are some words that might describe a rock's *luster*? (Answers might include: shiny, dull, and glassy.)

? What is one way to measure a rock's *size*? (Use a ruler to measure the longest side in centimeters.)

Discuss how observations of size such as big or small are not scientific observations because they are not exact. Using measurements to describe the size of a rock is more scientific. Then have students make careful observations of their rocks and complete their data tables.

Next, have students think about the unique properties of their rocks and fill in the cloze sentence, "I found a rock that would be good for _____ because it is _____."

Rock Sorting

This activity is a fun way to show that rocks can be identified by their unique properties. The object is to end with one student standing, holding his or her rock. Collect all of the I Found a Rock student pages. Randomly select one from the stack, but don't let students see it. Have all the students stand, holding their rocks. Then read the first observation on the page, for example, "I found a rock, and it is gray. If your rock is gray, stay standing." Students whose rocks are not gray should sit. Then read the second observation on the page, for example, "I found a rock, and it is smooth. If your rock is smooth, stay standing." Students whose rocks are not smooth should sit. Continue reading the observations, including the cloze sentence at the bottom, until only one student is standing. Repeat the process with several more student pages.

Measuring a rock

elaborate

Rocks: Hard, Soft, Smooth, and Rough Read Aloud and Rock Identification

Form groups of about five students. Give each student in a group a hand lens and one of the following rocks: obsidian, granite, sandstone, limestone, or marble. Have each student observe his or her rock and compare it to the other rocks in their group. Ask

? Are the five rocks all the same kind of rock? (no)

? How are they different? (They have different properties: shape, size, color, texture, and luster.)

? Is it possible to look at a rock and tell what kind of rock it is? (Answers will vary; the following activity will help students understand how geologists identify rocks by their properties.)

Next, tell students that the picture book *Rocks: Hard, Soft, Smooth, and Rough* can give them clues

about their rock's identity. Each one of the rocks they have been observing is described in the book. As you read the book aloud, stop after reading each rock description and ask students to hold up their rock if they think it is the one being described. After reading, use the rocks chart on page 21 to help students identify their rocks correctly. Explain that many different kinds of scientists use these kinds of charts, also called keys, to identify unknown objects.

After reading, use the following questions to help students understand how size and shape might not be good properties to use to identify rocks. Ask

? What properties did you use to identify your rock? (Answers might include: color, texture, luster, swirls, stripes, or specks.)

? Were you able to identify your rock based on its size or shape alone? (no)

? Why might size not be a good property to use to identify a rock? (Rocks are all different sizes depending on how they formed or broke apart from larger rocks—for example, a piece of granite can be any size.)

? Why might shape not be a good property to use to identify rocks? (For the same reason as in the previous question—for example, a piece of granite can be any shape.)

? What are the basic building blocks of rocks called? (minerals)

Next, have students use hand lenses to see if they can find any specks, crystals, swirls, or stripes in their rock samples. These are the minerals that make up their rocks. Some rocks are made of a single mineral, but most are made of several minerals. (A student who is observing a very fine-grained rock may not be able to see any minerals. Geologists often use special microscopes to look at very thin slices of rocks so that they can determine mineral content and thus rock type.)

Then ask

? What are the three main types of rocks you learned about in the book? (igneous, sedimentary, and metamorphic)

? How are rocks classified into these three

groups? (Rocks are classified based upon how they are formed.)

? How do scientists identify unknown rocks? (They can observe their properties and use a key.)

evaluate

Pet Rock Posters

Ask students if they have ever heard of a Pet Rock. Explain that way back in 1975, a businessman in California came up with the idea of selling rocks as pets. He considered dogs, cats, and birds too messy and expensive to keep, and instead advertised his Pet Rock as the ideal pet. The Pet Rock was packaged in a box that looked like a pet carrying case, and it even came with a "Pet Rock Training Manual." Topics included "How to Make Your Pet Rock Roll Over and Play Dead" and "How to House-Train Your Pet Rock." Believe it or not, the Pet Rock became a huge hit and the salesman became rich. (Optional: Show students a photo of the original Pet Rock and its packaging.)

Ask

? Would you have bought your own Pet Rock if you lived in the 1970s?

? Why do you think this businessman was able to sell so many Pet Rocks? (Answers might include: He had an original idea, and he used creative packaging and advertising.)

? What are some ways that advertisements help sell products? (Answers might include: They describe them, they make them sound useful, and they make them seem fun.)

Pass out the Pet Rock Poster Rubric and challenge students to create an advertisement for a Pet Rock. You may want to have them use either their own special rock or the rock they identified using the book *Rocks: Hard, Soft, Smooth, and Rough*. Have them give their rock a clever name ("Dusty," "Rocky," and "Cliff" come to mind!) and then design an ad to sell the rock. The advertisement should show what they have learned about properties of rocks, including

Using the Rock Identification Chart

- **4** Points: A detailed description of the Pet Rock's properties (including size, color, texture, and luster)

- **3** Points: A labeled, detailed, full-color drawing of the Pet Rock showing its unique features

- **2** Points: Two suggested uses for the Pet Rock based on its properties

- **1** Point: A creative statement to make people want to buy the Pet Rock

- Extra Credit: A poem, song, rap, jingle, drawing of the rock's packaging, or training tips for the Pet Rock

Have students share their advertisements with the rest of the class or have a gallery walk.

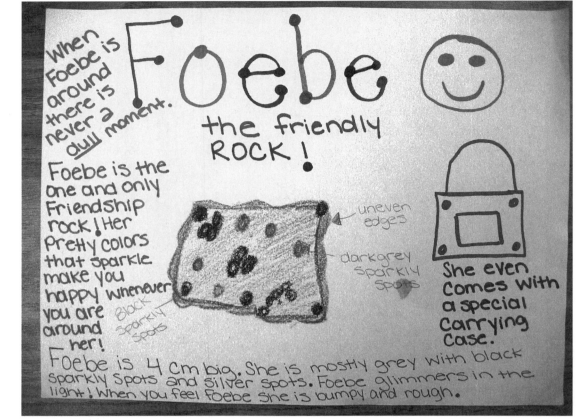

Pet rock advertisement

Inquiry Place

Have students brainstorm testable or researchable questions about rocks, such as

? How many ways can these rocks be sorted?

? How can you measure the volume of a rock?

? Which rock is the most common in our state?

? What can _____ be used for?

? Where is _____ found?

Then have students select a question to investigate or research as a class, or have groups of students vote on the question they want to investigate or research as a team. After they make predictions, have them design an experiment to test their predictions. Students can present their findings at a poster session or gallery walk.

More Books to Read

Baylor, B. 1985. *Everybody needs a rock.* New York: Aladdin.

Summary: Everybody needs a rock—at least that's the way this particular rock hound feels about it in presenting her own highly individualistic rules for finding just the right rock for you.

Gans, R. 1984. *Let's go rock collecting.* New York: Harper Collins Publishers.

Summary: This entry in the *Let's-Read-and-Find-Out Science* series describes the formation and characteristics of igneous, sedimentary, and metamorphic rocks and how to recognize and collect them.

Hooper, M. 1996. *The pebble in my pocket.* New York: Viking Juvenile.

Summary: A girl finds a pebble on the ground and wonders where it came from. The answer unfolds through scientifically accurate text, colorful illustrations, and a helpful timeline that follows its long journey from the inside of a volcano to the day the girl picks it up off the ground.

Hurst, C.O. *Rocks in his head.* New York: Greenwillow Books.

Summary: Based upon true events in the life of the author's father, this book tells the story of how a young man's lifelong love of rocks eventually lead him to work at a science museum.

Website

Rock Hounds: Learn how to collect rocks safely, find out how rocks are formed, take a quiz, and more.

http://sln.fi.edu/fellows/payton/rocks/index2.html

Rock Hound's Name: _____

I Found a Rock

Color What colors or patterns does it have?	Texture How does it feel?	Luster How shiny or dull is it?	Size What is the longest length in cm?

Labeled Drawing of My Rock

I found a rock that would be good for _____

because it is _____ .

Pet Rock

Advertising Poster Rubric

Your Name: _____ *Pet Rock's Name:*_____

Your poster includes:

4 Points: A detailed description of the Pet Rock's properties (including size, color, texture, and luster)

    4   3   2   1   0

3 Points: A labeled, detailed, full-color drawing of the Pet Rock showing its unique features

         3   2   1   0

2 Points: Two suggested uses for the Pet Rock based on its properties

              2   1   0

1 Point: A creative statement to make people want to buy the Pet Rock

                    1   0

Extra Credit: Your poster includes a poem, song, rap, jingle, drawing of the rock's packaging, or "training tips" for the Pet Rock

                    1   0

Total Points_____/10

Comments: _____

Sunshine on My Shoulders

Description

Learners conduct simple experiments to find out what makes "mystery beads" change color outdoors. They discover that invisible rays of UV (ultraviolet) light cause the beads to change color. After discussing the harmful effects of too much sun on their skin, learners test the effectiveness of sunscreen on UV beads and then illustrate a sun safety poster.

Suggested Grade Levels: K–2

Lesson Objectives Connecting to the Standards

Content Standard A: Scientific Inquiry

- Ask a question about objects, organisms, and events in the environment.
- Design and conduct simple experiments to answer questions.
- Use data to construct reasonable explanations.

Content Standard D: Earth and Space Science

- Understand that the Sun provides the light and heat necessary to maintain the temperature of the Earth.

Featured Picture Book

Title	*Sunshine on My Shoulders*
Author	John Denver
Illustrator	Christopher Canyon
Publisher	Dawn Publications
Year	2003
Genre	Story
Summary	A picture book adaptation of John Denver's song "Sunshine on My Shoulders," which celebrates the simple things in life such as sunshine, nature, and loving relationships.

Time Needed

This lesson will take several class periods. Suggested scheduling is as follows:

Day 1: **Engage/Explore** with Mystery Beads.

Day 2: **Explain** with Sun Safety Article.

Day 3: **Elaborate** with *Sunshine on My Shoulders* read aloud and Sunscreen Test.

Day 4: **Evaluate** with Sun Safety Tips.

Materials

UV beads (5 per student)

Pipe cleaners or long twist ties (1 per student)

White and colored crayons or pencils

1 bottle of sunscreen (at least SPF 30)

Student Pages

Mystery Beads

Sun Safety article

Sun Safety Tip

UV beads are available from

Sheridan WorldWise

www.classroomgoodies.com

1-800-433-6259

Educational Innovations, Inc.

1-888-912-7474

www.teachersource.com

Background

The National Science Education Standards suggest that, in grades K–4, students acquire the essential understanding that the Sun provides the heat and light necessary to maintain the temperature of the Earth. In this lesson, students discuss the benefits of sunshine, such as heat and light, as well as the risks that come with sunlight. It is important for children to learn at a very young age that overexposure to sunlight can damage their skin. Through observing UV (ultraviolet) beads, students discover that the Sun emits UV light. The plastic beads used in this lesson contain a chemical that changes color when exposed to ultraviolet (UV) radiation. The colors that develop depend on the wavelength of the UV radiation. Skin is also a detector of UV light. When bare skin is exposed to sunlight, it becomes darker (suntan) or redder (sunburn). These responses by the skin are a signal that the cells under the skin are being assaulted by UV radiation from the Sun. UV radiation breaks chemical bonds in skin tissue and over prolonged exposure, skin may wrinkle or skin cancer may develop.

SAFETY — When taking students outdoors, remind them to never look directly at the Sun. Looking at the Sun can damage their eyes.

ngage/Explore

Mystery Beads

Have each student make a mystery bead bracelet by stringing five UV beads on a pipe cleaner or long twist tie. Have them twist a bracelet on their wrists. Tell students that they are wearing a special bracelet made with mystery beads. Pass out the Mystery Beads student page and have students color the indoors picture with a white crayon showing that the beads are white inside the classroom. When they finish coloring, take them outside to a sunny spot and have them pay close attention to any changes in their bracelets.

The beads will change color when in sunlight, but *do not tell students the reason for the change yet.* Have students come back into the classroom and promptly color the outdoors picture of the beads before they change back to white.

When students are finished coloring, have them brainstorm explanations as to why the beads changed color outside. Write these explanations on chart paper or the board. Possible student responses include

● The hot or cold air made them change.

● The water in the air made them change.

● The wind made them change.

● The light made them change.

Observing the "mystery beads"

Ask students

? How can we find out for sure what causes the beads to change color outside? (We can test our explanations one by one.)

Depending on time or availability of materials, these tests can be done as a group demonstration or in student teams. Each class will have different explanations to test, but following are some examples of tests for some common student explanations. When an explanation is tested and proved to be the wrong explanation, cross it off the list.

Explanation	Recommended Test
The heat made the beads change.	Hold the bracelet tight in your hand to warm it up.
The cold made the beads change.	Put an ice cube on the beads.
The water in the air made the beads change.	Dip the beads in water or spray them with water.
The wind made the beads change.	Hold the beads in front of a fan.
The bright light made them change.	Hold the beads under a bright lamp.
Something special about the sunlight made them change.	Hold the beads in the sunlight.

explain

Sun Safety Article

After all of the tests, the only explanation that will remain is "Something special about the sunlight made them change." Tell students that you have an article that will help them find out more about sunlight.

📖 Determining Importance/Pairs Read

Give each student a copy of the Sun Safety article. Read the article together or pairs read, and ask students to watch and listen for any clues about what is special about sunlight that made the beads change color.

After reading, ask

? What caused the mystery beads to change color outdoors?

As a class, write a sentence answering this question on the board. For example, "The beads changed color outdoors because of the UV light from the Sun."

Tell students that when scientists come up with answers, they always share their evidence or proof that the answer is the right one. Ask students,

"Sunshine on My Shoulders" sing-along

? What is our evidence (proof) that the UV light caused the beads to change color? (Our evidence is that we tried the other explanations and they did not work. We also read that UV light from the Sun causes UV beads to change color.)

elaborate

Sunshine on My Shoulders
Read Aloud and
Sunscreen Test

Making Connections:
Text to Self

Show students the cover of the book *Sunshine on My Shoulders*. Ask

? How does it make you feel to have the Sun shining on your shoulders?

Tell students that the book you will read to them is actually a song written by John Denver. He believed that the most wonderful things in life are simple and free, like sunshine and friendship. An artist by the name of Christopher Canyon created pictures to go along with the song. You may want to play the CD (enclosed in the hardback version of the book) or sing the lyrics as you show students the pictures. Invite students to sing along.

Ask students

? What are some good things about sunshine? (Answers might include: It makes plants grow, or it makes the earth warm.)

? What are some harmful things about sunshine? (It can burn your skin and damage your eyes.)

? Have you ever had a sunburn? How did it make you feel?

? What are some ways to protect your skin and eyes from sunshine? (Answers might include: hats, sunglasses, visors, and sunscreen.)

Tell students that UV light is the light that burns our skin when we are in the sun. We wear sunscreen to protect our skin from UV light. Ask students

? What do you think would happen if we put sunscreen on the UV beads?

Test this question by generously coating five UV beads with sunscreen (SPF 30 or higher). Take those beads and five beads without sunscreen into the sun and let students observe them. The UV beads with sunscreen will appear very faint in color compared to the bright colors of the UV beads without sunscreen. Ask students

? How does the color of the UV beads with sunscreen compare to the UV beads without sunscreen? (The UV beads with sunscreen are lighter in color.)

? Why do you think they look different? (The sunscreen blocks UV light.)

? Do you think the sunscreen blocks all UV light? (No, because the beads with sunscreen did turn color slightly. The sunscreen blocks only some UV light.)

Observing mystery beads with and without sunscreen

evaluate

Sun Safety Tips

Give each student a copy of the Sun Safety Tip student page. Have them fill in the blank in the sentence, "When you are in the sun, wear _____. It helps block UV light." Students can fill in the blank with words such as *sunscreen, a hat, sunglasses, long sleeves.* Then have students draw a picture of themselves wearing the protective items.

A sun safety tip

Inquiry Place

Have students brainstorm testable questions such as

? Do sunglasses block UV light?

? Does fabric block UV light?

? How do UV beads covered with SPF 8, SPF 15, SPF 30, or SPF 45 compare in color outside?

? Do clouds block UV light from the Sun?

Have students select a question to investigate as a class, or have groups of students vote on the question they want to investigate as a team. After they make their predictions, they can design an experiment to test their predictions. Students can present their findings at a poster session or gallery walk.

More Books to Read

Asch, F. 2000. *The Sun is my favorite star.* New York: Gulliver Books.

Summary: This book celebrates a child's love for the Sun and explores the wondrous ways in which it helps the Earth and the life upon it.

Branley, F. 1986. *What makes day and night?* New York: HarperTrophy.

Summary: This Let's-Read-and-Find-Out Science book provides a simple explanation of how the rotation of the Earth causes day and night. A student activity for modeling Earth's rotation is included.

Fowler, A. 1991. *The Sun is always shining somewhere.* Chicago: Children's Press.

Summary: This Rookie Read-About Science book gives simple examples of how the Sun is important to life on Earth, explains that the Sun is a star, and provides a simple student activity to show that Earth's rotation causes day and night and that the Sun is always shining.

Rau, D. M. 2006. *Hot and bright: A book about the Sun.* Minneapolis: Picture Window Books.

Summary: Simple text, colorful illustrations, and fun facts explain how the Sun moves, how it helps the Earth, how to protect yourself from harmful rays, and more. Includes simple experiments, table of contents, a glossary, and a website with links to other safe, fun websites related to the book's content.

Sherman, J. 2003. *Sunshine: A book about sunlight*. Minneapolis: Picture Window Books.

Summary: Sunrises, sunsets, rainbows—all these things come from the Sun. Sunshine also gives us light, warmth, and food. In this book about sunlight, readers find out how the Sun creates all of our weather. Includes simple experiments, table of contents, a glossary, and a website with links to other safe, fun websites related to the book's content.

Tomecek, S. 2001. *Sun*. Washington, DC: National Geographic Society.

Summary: This information-packed book explains that the Sun is a star and that Earth's rotation causes the Sun to appear to move across the sky each day. Readers also learn how far the Sun is from Earth, what it is made of, how hot it is, and many other facts about the Sun.

Name: _____

Mystery Beads

Indoors
1. Color the mystery beads the way they look indoors.

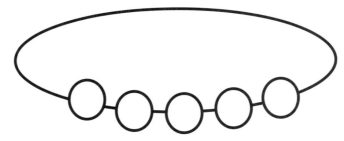

Outdoors
2. Color the mystery beads the way they look outdoors.

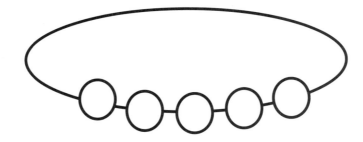

3. What do you think caused the mystery beads to change?

_____

_____

_____

Sun Safety

Have you ever been outside in the sun for a long time? Did your skin turn red or hurt when touched? You probably had a sunburn!

The Sun gives us more than just light we can see. It also gives off rays that are invisible. Some of these rays are called **ultraviolet** (UV) light rays.

UV light from the Sun causes sunburns and can also hurt your eyes. You can protect your skin and eyes from the Sun by wearing sunscreen, sunglasses, and a hat.

Special beads called UV beads will change color when put in UV light. UV beads are white in the light of a lamp and show their colors when they are in sunlight. The brighter the colors of UV beads, the stronger the UV light.

Name: _____

Sun Safety Tip

When you are in the Sun, wear _____.
It helps block UV light.

Stargazers

Description

Learners are introduced to the idea of looking up in wonder at the night sky and asking questions about what they see. They observe the stars and record their observations and wonderings. They also learn about constellations—the names, pictures, and stories that people have invented to explain the patterns of stars in the night sky.

Suggested Grade Levels: 2–4

Lesson Objectives Connecting to the Standards

Content Standard A: Scientific Inquiry

- Ask a question about objects, organisms, and events in the environment.

Content Standard D: Earth and Space Science

- Understand that objects in the sky have properties, locations, and movements that can be observed and described.

Featured Picture Books

Title	*When I Heard the Learn'd Astronomer*	*Spots of Light: A Book About Stars*
Author	Walt Whitman	Dana Meachen Rau
Illustrator	Loren Long	Denise Shea
Publisher	Simon & Schuster	Picture Window Books
Year	2004	2006
Genre	Story	Non-narrative Information
Summary	Walt Whitman's beautiful words are recast to tell the story of a boy's fascination with the heavens.	Simple text and digitally generated illustrations explains the birth of stars, star colors, constellations, and galaxies.

Time Needed

This lesson will take several class periods. Suggested scheduling is as follows:

Two Weeks Prior: **Engage** with *When I Heard the Learn'd Astronomer* Picture Walk and read aloud, and **Explore** at home with Stargazing.

Day 1: **Explain** with *Spots of Light* read aloud.

Day 2: **Elaborate** with Constellations and Pictures in the Sky.

Day 3: **Evaluate** with Stargazers Lift-the-Flap Book.

Materials

Suggested Materials for Stargazing at home:

> Flashlights
>
> Clipboard or notebook
>
> Pen or pencil

O-W-L chart (with teacher observations and wonderings filled out in advance)

Large sticky notes (about 3 × 5)

Crayons or markers

Canis Major overhead

Student Pages

Stargazing Take-Home Page

Pictures in the Sky

My Constellation

Stargazers Lift-the-Flap Book

Background

Children are naturally fascinated by the objects in the night sky. By observing the night sky regularly, children will begin to notice changes and the patterns in these changes. The National Science Education Standards suggest that, for young children, understandings about objects in the sky be limited to observations, descriptions, and finding patterns. Attempting to extend this understanding into explanations that involve models is not recommended for young children. Instead, the Standards suggest that the emphasis in grades K–4 be on encouraging students to develop observation and description skills. Young students should be invited to draw and talk about what they see and think when they are observing the sky. In the Explore phase of this lesson, students observe the night sky with an adult helper and record their observations and wonderings. A useful internet resource for stargazing is Star Date, located at *www.stardate.org*. The website includes stargazing tips for beginners, a sky almanac, information on constellations, and much more. Another option for stargazing is to have a star party at which the teachers, students, and parents meet to look at the stars together. This can be held in the schoolyard, an observatory, or local park. Contact a local astronomy club or observatory center to assist with setting up a star party for your class. In the Explain phase of the lesson, students learn from a nonfiction read aloud that stars are round, different sizes, and different colors depending on how hot

they are, and that there are more stars than anyone could ever count. The Elaborate phase of this lesson deals with understanding that *constellations* are patterns of stars that people have imagined as various figures, animals, or objects. Over thousands of years, many civilizations have tried to explain the patterns of stars in terms of their own cultures. These patterns, or constellations, include such pictures as a bear, a dog, a lion, and a hunter. To try to explain these patterns, people gave them names and made up stories about them. The International Astronomical Union recognizes 88 constellations. It is important for students to understand that the names and stories behind the constellations were made up in the imaginations of people. The night sky is what we see when our side of the Earth faces away from the Sun. Due to the Earth's revolution around the Sun, we look out on different parts of the universe at night. This is why we see different constellations at different times of the year.

engage

When I Heard the Learn'd Astronomer Picture Walk and Read Aloud

Picture Walk

Introduce the author of *When I Heard the Learn'd Astronomer*. This book is based on a poem by the famous American poet Walt Whitman, who died more than a hundred years ago. Then introduce the illustrator, Loren Long. He has illustrated many picture books for children, including a version of *The Little Engine That Could*. More information about this illustrator can be found at *www.lorenlong.com*.

Before reading the book, take students on a picture walk of *When I Heard the Learn'd Astronomer* by showing them the cover and browsing through the pages in order. Encourage students to talk about what they see, what may be happening in each illustration, and how the pictures come together to make a story. Some useful questions to model are:

? From looking at the cover, what do you think this book is about?

? What do you see?

? What do you think is happening?

? What do you think will happen next?

? What are you curious to know more about in the story?

Now read the book aloud to students. Some of the vocabulary in the Walt Whitman poem may be difficult for students. You may want to focus more on the illustrations as a way of making meaning from the book.

Synthesizing

Invite students to talk with a partner or a small group about what they think the story means. Turn back to the pages at the end of the book showing the boy sitting in the lecture room and looking up at the stars and ask

? How do you think the boy was feeling when he was sitting in the lecture room? (Answers might include: bored or tired.)

? How do you think the boy was feeling when he was looking up at the stars at the end of the book? (Answers might include: happy, thoughtful, or full of wonder.)

? Do you think it would be more enjoyable to listen to someone talk about the stars or actually look at the stars yourself?

The process of synthesizing may be challenging for some students. It may help for you to model your own synthesis of the story's meaning. For example, you could say

> When I took a picture walk through the book, it looked to me like the boy was going somewhere that he wasn't very happy about. It looked like he was in a museum and then

Taking a picture walk

in a lecture hall. He looked really bored sitting in his seat. I think he left because he was bored. When he got outside and looked at the stars he looked interested and happy. When I read the words of the book, I realized that the boy was going to hear an astronomer. I know that an astronomer is someone who studies stars and planets and other things like that. The astronomer was talking in the lecture hall about charts, diagrams, and how to add, divide and measure. The boy was tired and sick of hearing the lecture, so he walked out.

When he looked up at the stars at the end of the book, he seemed happier. It looked like he was wondering about the night sky. I realized this book is not just about getting bored by a speech and wandering away from it. I think the message is this—when you are learning about the natural world, you need to be sure to keep a sense of wonder about it.

I read about the author, Walt Whitman, on the back flap of the book. He was a teacher for a short time, and he believed in encouraging his students to think out loud rather than just memorize facts. I think his message is that when you learn about the night sky, you shouldn't just memorize facts and numbers. You should also go outside and look up "in perfect silence at the stars."

explore

Stargazing

Ask students

? Have you ever "looked up in perfect silence at the stars" like the boy in the book?

? What can you see in the night sky? (Answers might include: airplanes, spots of light, stars, clouds, planets, and the Moon.)

Give students the Stargazing Take Home page. Tell them that sometime in the next few weeks, their homework is to go outside with an adult helper, look at the night sky, and complete an Observations and Wonderings chart. The Stargazing student page includes a brief note to parents, items to bring, tips for stargazing, and an Observations and Wonderings chart.

explain

Spots of Light Read Aloud

In advance, make a large O-W-L (**O**bservations, **W**onderings, **L**earnings) chart with the observations and wonderings listed on the chart and post it in the classroom. Tell students that you have some observations and wonderings about stars. Read each observation aloud to the students. Then explain that your observations led you to

some wonderings about stars. Read each wondering aloud to students. The questions on this chart can all be answered in the book *Spots of Light: A Book About Stars.*

Before you read the book, tell students that you would like them to share their own observations and wonderings from the Stargazing Take-Home page. Put students in groups of three or four and have them discuss their observations and wonderings. After a few minutes of discussion, ask each group to choose one observation and write it on a large sticky note and then choose one wondering and write it on another large sticky note. Invite each group to read their sticky note observations and wonderings aloud and then post them on the whole-class O-W-L. Tell students that as questions are answered throughout the lesson, they can add to the learnings column of the whole-class O-W-L chart.

O Observations	W Wonderings	L Learnings
Stars look like balls of light.	What shape are stars?	
Some stars seem brighter than others.	Are all stars the same size?	
Some stars look different colors.	What color are stars?	
Some stars make patterns in the sky.	What are these patterns called?	
I can see a lot of stars.	How many stars are there?	

Using Features of Nonfiction

Show students the cover of the book *Spots of Light: A Book About Stars,* and tell them that this book might answer some of these questions about stars. Begin by reading the table of contents aloud to the students. Explain that by reading the table of contents in a nonfiction book, you can find out quickly what the book is going to be about. Ask

? Do you think *Spots of Light* will be a good resource for answering questions about stars? Why or why not?

Determining Importance

Note: Skip pages 20–21 in Spots of Light, because it may give students misconceptions that stars have feelings or that stars are close together.

As you read *Spots of Light* aloud to students, have them signal when they hear the answer to one of the questions from the wonderings column of the O-W-L chart. After reading, add to the learnings column of the chart at the bottom of the page.

As you find information that answers any of the questions on the whole-class O-W-L chart, be sure to add the answers to the learnings column. This chart can be posted in the room and added to throughout the entire lesson.

elaborate
Constellations

Explain that people who lived long ago looked at the stars often. They imagined that groups of stars made pictures, kind of like dot-to-dot pictures. These patterns or pictures are called *constellations.* People gave names to the pictures that have been passed on from generation to generation and are still used today. People also made up stories about the constellations.

Pairs Read

Pass out the Pictures in the Sky student page, and pairs read the name of each constellation and its story. Students will probably notice that the shape of each constellation does not look exactly like the person or animal for which it was named. People used a lot of imagination to come up with the names and stories for constellations. Mention that students will not be able to see these constellations on all nights because of the Earth's revolution around the Sun.

Next, give students crayons or markers and pass out the My Constellation student page. Invite

O	W	L
Observations	**Wonderings**	**Learnings**
Stars look like balls of light.	What shape are stars?	Stars are round.
Some stars seem brighter than others.	Are all stars the same size?	Stars are different sizes.
Some stars look different colors.	What color are stars?	Stars can be different colors.
Some stars make patterns in the sky.	What are star patterns called?	Star patterns are called constellations.
I can see a lot of stars.	How many stars are there?	There are more stars in the sky than anyone could ever count.

Using the table of contents

students to use their imaginations to create their own constellations and stories out of the star pattern on the page.

When students are finished, have them share their constellations with the class. Ask

? Why do you think we all have different pictures, names, and stories when we all started with the same pattern of stars? (We all used our own imaginations to make up the pictures, names, and stories.)

Next, show students the Canis Major overhead. Explain that long ago, the Greeks saw the same pattern of stars in the sky and imagined that it made a picture of a dog. They called this constellation Canis Major, or the Great Dog. Read the brief story of Canis Major on the overhead, and then ask

? How does your picture compare to Canis Major?

? Is there one right way to connect the dots and

imagine a picture? (No, everyone imagines something different.)

Explain that even though the names, pictures, and stories associated with constellations are all made up, astronomers still use the constellations to find certain stars in the night sky. Tell them that there are 88 constellations that are recognized and used by astronomers.

evaluate

Stargazers Lift-the-Flap Book

Tell students that they are going to have an opportunity to show what they have learned about stars. Give each student a copy of the Stargazers Lift-the-Flap Book. Have them fold each page on the dotted line and then staple the pages together. For each question, students can write and draw their answer.

Inquiry Place

Have students brainstorm testable and researchable questions such as

? Do constellations look different at different times of night?

? Do constellations look different at different times of the year?

? What is the closest star to Earth?

Have students select a question to investigate as a class, or have groups of students vote on the question they want to investigate as teams. After they make their predictions, they can design an experiment to test their predictions. Students can present their findings at a poster session or gallery walk.

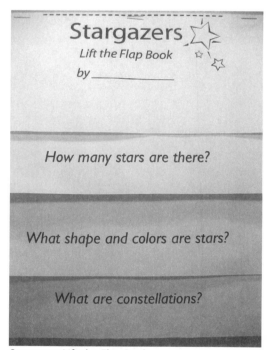

Stargazers Lift-the-Flap Book

More Books to Read

Asch, F. 2000. *The Sun is my favorite star*. New York: Gulliver Books, Harcourt.

Summary: This book celebrates a child's love of the Sun and the wondrous ways in which it helps the Earth and the life upon it.

Branley, F.M. 1991. *The big dipper*. New York: Harper-Trophy.

Summary: This Let's-Read-and-Find-Out-Science book explains basic facts about the Big Dipper, including which stars make up the constellation, how its position changes in the sky, and how it points to the North Star.

Branley, F.M. 1981. *The sky is full of stars*. New York, HarperTrophy.

Summary: This Let's-Read-and-Find-Out-Science book explains how to view stars and ways to locate star pictures, known as constellations, throughout the year.

Gibbons, G. 1992. *Stargazers*. New York: Holiday House.

Summary: This book tells what stars are, why they twinkle, how constellations were named, how telescopes are used to study stars, and more.

Malone, P. 1997. *Star shapes*. San Francisco: Chronicle Books.

Summary: Rhyming text and beautiful paintings highlighted with silver stars describe some of the animal constellations that can be seen in the night sky.

Thompson, C.E. 1989. *Glow in the dark constellations: A field guide for young stargazers*. New York: Grosset and Dunlap.

Summary: Shows pictures of 10 constellations printed in glow-in-the-dark ink. Includes the myths about the constellations as well as when and where to find them in the night sky.

Tomecek, S. 2003. *Stars*. Washington, DC: National Geographic.

Summary: Introduces stars and what they are made of, how they shine, their positions in relation to Earth, and more.

Canis Major, the Great Dog

The Story of Canis Major

The Great Dog belongs to Orion the hunter. He follows his master across the sky, forever chasing Lepus, the rabbit. Canis Major could run incredibly fast. He won a race against a fox that was the fastest creature in the world. The dog was placed in the sky to celebrate the victory.

Name: _____

Stargazing
Take-Home Page

Dear Parent,

At school, we are studying the stars. Your child's homework assignment is to go outside with an adult helper and observe the night sky. As you look up at the sky with your child, help him or her record "Observations" and "Wonderings" (questions) on the Stargazing Chart. Below is a list of items to take outside with you and some tips on stargazing. The purpose of this assignment is to give your child the opportunity to observe the stars firsthand and wonder about them. In class, we will be building on this experience by learning more about stars and constellations.

This assignment is due by _____ .

Items to Take Outside:

Flashlights
Clipboard or notebook
Stargazing Observations and Wonderings Chart
Pen or pencil

Stargazing Tips:

1. Using a flashlight to light the way, find a place to stargaze that is away from bright lights.

2. Turn your flashlight off, and allow at least 10 minutes for your eyes to adust to the darkness. When you first look at the sky, the stars may all seem alike to you, but take your time to look for a while and you may begin to see differences.

3. After observing for a while, turn on your flashlight and write down some observations and wonderings on the chart.

Child: _____

Adult Helper: _____

Date: _____ Time: _____

Stargazing
Take-Home Page cont.

Observations	Wonderings

Name: _____

Pictures in the Sky

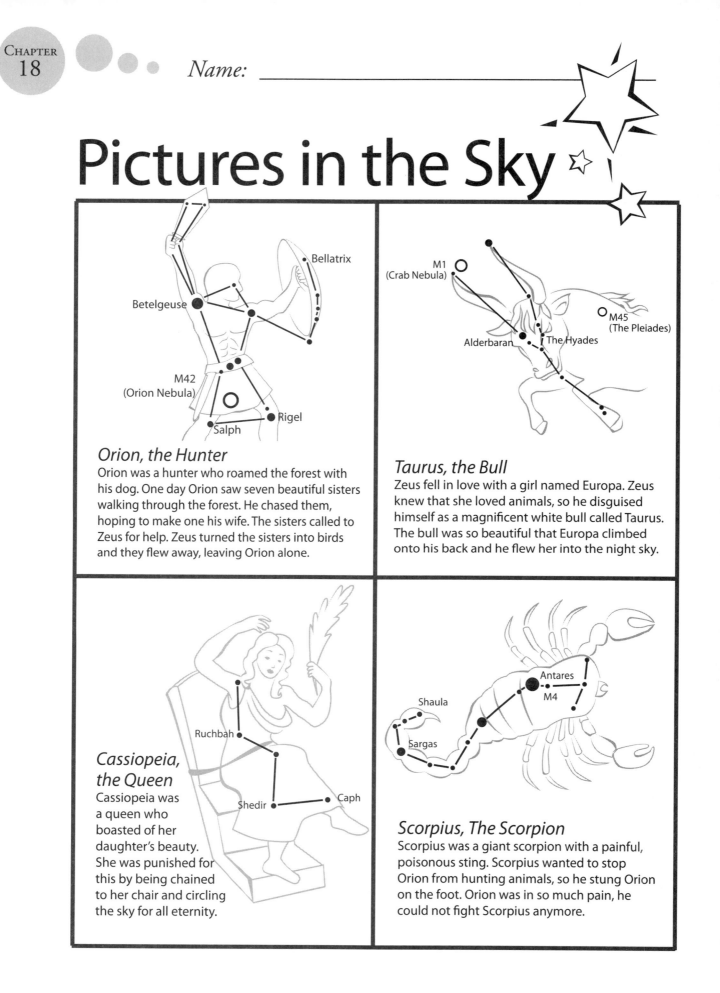

Orion, the Hunter

Orion was a hunter who roamed the forest with his dog. One day Orion saw seven beautiful sisters walking through the forest. He chased them, hoping to make one his wife. The sisters called to Zeus for help. Zeus turned the sisters into birds and they flew away, leaving Orion alone.

Taurus, the Bull

Zeus fell in love with a girl named Europa. Zeus knew that she loved animals, so he disguised himself as a magnificent white bull called Taurus. The bull was so beautiful that Europa climbed onto his back and he flew her into the night sky.

Cassiopeia, the Queen

Cassiopeia was a queen who boasted of her daughter's beauty. She was punished for this by being chained to her chair and circling the sky for all eternity.

Scorpius, The Scorpion

Scorpius was a giant scorpion with a painful, poisonous sting. Scorpius wanted to stop Orion from hunting animals, so he stung Orion on the foot. Orion was in so much pain, he could not fight Scorpius anymore.

Name: _____

My Constellation

Use your imagination to create a picture out of the pattern of stars below. Choose a name for your constellation and write a story about it.

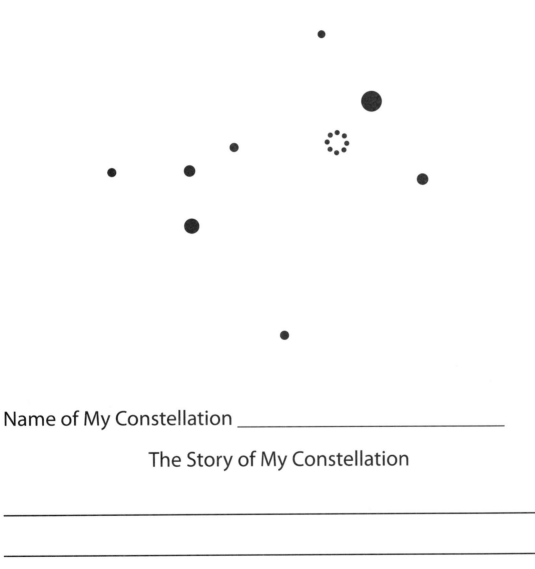

Name of My Constellation _____

The Story of My Constellation

_____

_____

_____

_____

_____

Stargazers

Lift the Flap Book

by _____

- -

How many stars are there?

- -

What shape and colors are stars?

What are constellations?

Imaginative Inventions

Description

Learners explore the invention process by learning about inventions throughout history and how inventions fill needs or wants, by improving existing inventions, and by keeping a toy invention journal. They further their understandings of the risks and benefits of inventions by testing toys and comparing the fun rating and the safety rating of each toy.

Suggested Grade Levels: 2–4

Lesson Objectives Connecting to the Standards

Content Standard E: Science and Technology Abilities of Technological Design

● Identify a simple problem, and identify a specific task and solution related to the problem.

● Propose a solution to make something work better.

● Evaluate a product or design made by themselves or others.

Content Standard E: Science and Technology Understanding About Science and Technology

● Understand that people have always had problems and invented tools and techniques to solve problems.

● Understand that trying to determine the effects of solutions helps people avoid some new problems.

Featured Picture Books

Title	*Imaginative Inventions*	*Leo Cockroach, Toy Tester*
Author	Charise Mericle Harper	Kevin O'Malley
Illustrator	Charise Mericle Harper	Kevin O'Malley
Publisher	Little, Brown	Walker
Year	2001	1999
Genre	Dual Purpose	Story
Summary	The who, what, where, when, and why of roller skates, potato chips, marbles, and pie told in rhyming verse.	Leo Cockroach, who secretly tests toys for the bug-hating president of a toy company, seeks a job with the competitor across the street and finds himself worse off than before.

Time Needed

This lesson will take several class periods. Suggested scheduling is as follows:

Day 1: **Engage** with *Imaginative Inventions* read aloud, and **Explore/Explain** with Improve an Invention.

Day 2: **Elaborate** with *Leo Cockroach, Toy Tester* read aloud and Toy Testing.

Day 3 and Beyond: **Evaluate** with Toy Invention Journal and Advertising Poster.

Materials

One standard Frisbee flying disc

One pie tin

Several "new and improved" versions of flying discs (or pictures of them) such as a foam flying disc, the Glow in the Dark Frisbee, or an Aerobie ring

Other examples of inventions from the book *Imaginative Inventions*, such as eyeglasses, high-heeled shoes, roller skates, flat-bottomed paper bags, marbles, and piggy banks

2 types of inexpensive simple toys to test (1 of each per student or pair), such as blow ball pipes, finger traps, jumping frogs

Fun Ratings overhead

Safety Ratings overhead

Student Pages

My Improved Invention

Toy Testing

My Toy Invention Journal

> **Inexpensive toys are available from**
> *www.orientaltrading.com*
> and
> WorldWise Sheridan
> *www.classroomgoodies.com*

Background

The design process in technology is the parallel to inquiry in science. In scientific inquiry, students explore ideas and propose explanations about the natural world, whereas in technological design students identify a problem or need, design a solution, implement a solution, evaluate a product or design, and communicate the design process. In grades K–4, the standards suggest studying familiar inventions to determine function and to identify problems solved, materials used, and how well the product does what it is supposed to do. The purpose of this lesson is to encourage students' creativity, imagination, and problem-solving skills with activities that focus on the technological design process.

In today's fast-growing, highly competitive global marketplace, innovative thinking is more important than ever. Technology involves using science to solve problems or meet needs, and the understanding of technology can be developed by challenging students to design a solution to solve a problem or invent something to meet a need. Simply stated, inventors try to solve problems. They think about peoples' *needs* and come up with a solution. Inventions don't have to be entirely new ideas. Sometimes they can be add-ons or improvements to existing inventions. All inventions have *benefits* (good things that result from using them) and *risks* (possible hazards that may result from using them). Inventors must be sure the benefits of their inventions outweigh the risks.

Engage

Imaginative Inventions
Read Aloud

📖 Making Connections: Text to World

Show the cover of the book, *Imaginative Inventions,* and introduce the author and illustrator. Ask

? What is an invention? (Something that is made to meet a need or solve a problem.)

? What is the difference between an invention and a discovery? (An invention is something that is created; a discovery is something that is found for the first time. For example, Ben Franklin discovered that lightning is electrical current, but he invented the lightning rod.)

? What do inventors do? (They think about people's needs or problems and come up with solutions.

Build connections to the author by reading the inside flap of the book about Charise Mericle Harper's favorite invention ("… muffins, which taste a lot like cake, but you get to eat them for breakfast!") Ask

? What do you think is the greatest thing ever invented? Turn and talk to a partner.

📖 Inferring: Stop and Jot

Select several of the inventions in the book to read about. As you read each two-page spread, leave out the name of the invention and instead say "this invention." Have students make inferences about the identity of each invention using clues from the text and illustrations. They can stop and jot their guesses on sticky notes as you read. After reading each description, reveal the name of the invention and then have students identify the need or want that the invention filled.

Explore/ Explain

Improve an Invention

Explain that instead of coming up with completely new inventions, inventors often think of ways to make an old one better. A good example of this involves the improvements made to a very popular toy, the Frisbee. Make a T-chart with the words *benefit* and *risk* on the board. Discuss that all inventions can have both benefits (good results) and risks (bad results) for people and the environment. Ask the following questions, and write the students' responses on the T-chart:

? What are the possible benefits of a Frisbee? (Answers might include: You can have fun with it and get exercise using it.)

? What are the possible risks of a Frisbee? (Answers might include: You could get hurt if hit by a Frisbee, you could lose it outside, creating litter and making Frisbees in factories could cause pollution.)

Comparing flying toys

Explain that the original Frisbee had a serious risk: It was made of a very hard plastic that could really hurt if it hit you! Inventors improved upon this by making it from a softer material so the Frisbee was less risky to use. Demonstrate the evolution of Frisbee design by showing students a pie tin as well as several "new and improved" versions of flying discs such as lightweight foam versions, the Glow-in-the-Dark Frisbee, or an Aerobie ring. (You may want to take students outside to test some of the improved versions and compare them to the original.) Ask

? How are the new and improved Frisbees more fun or useful than the original?

? What are the benefits of the new and improved Frisbees?

? What are the possible risks of the new and improved Frisbees?

Explain that inventors try to improve products by increasing their benefits and reducing their risks.

Brainstorming ways to improve an invention

Now go back to *Imaginative Inventions* and write the names of the other inventions from the book on the board. Provide examples of several of these for students to look at, such as eyeglasses, high-heeled shoes, roller skates, flat-bottomed paper bags, and marbles. Have each student or group choose one of the inventions from the book and brainstorm ways that the invention could be improved upon.

Pass out the My Improved Invention student page. Have students select one of their ideas for improving an invention, draw a labeled picture of it, and give it a clever or descriptive name. They should also explain how their improved invention is more fun or more useful than the original and describe its risks and benefits. Student directions for the My Improved Invention page are as follows:

1 Which invention would you like to improve?

2 Draw and label your improved invention in the box below, and give it a new name.

3 How is your improved invention more fun or useful than the original?

4 What are the benefits of your improved invention?

5 What are the risks of your improved invention?

elaborate

Leo Cockroach, Toy Tester
Read Aloud and Toy Testing

Introduce the author and illustrator of the book *Leo Cockroach, Toy Tester.* Kevin O'Malley first decided he wanted to illustrate children's books when he was in the fourth grade! He was in "time out" one day when he started reading *Where the Wild Things Are* by Maurice Sendak, and that book inspired him to write and illustrate humorous books for kids. (For more information on this author and illustrator, go to *http://mywebpages. comcast.net/komalley.*)

Thumbs down for "The Pointy Stick"

📖 Inferring

Ask students to look at the cover and title of the book and make an inference.

❓ What do you think this book is about?

❓ Do you think toy testing is a real job? (Toy companies have to test their toys for safety. Many toy companies also give children toys to test and observe their reactions to them.)

📖 Determining Importance

Before you begin reading the book, ask students to give a thumbs up for any toy in the book they think would be fun and safe and a thumbs down for any toy that they think would be boring or dangerous. Then read aloud *Leo Cockroach, Toy Tester*, making sure to read the name of each toy in the illustrations. After reading ask

❓ Why do companies need to test toys? (To see if the toys are both fun and safe.)

Discuss how most toys come with warning labels and/or directions for using the toy safely. Discuss the possible risks of various toys. For example, babies and young children often put things in their mouths. If a toy is too small, or contains small parts, it could be a choking hazard. If a toy is too sharp it could poke someone.

Then ask

❓ Would you like to be a toy tester?

Tell students that they are going to have a chance to be toy testers for Waddatoy Toys! Pass out the Toy Testing student page. On the board, write the name of one of the toys and label it "Toy A." Write the name of the other toy on the board and label it "Toy B." Give each student or pair of students both toys to test. They will be testing how much fun and how safe each toy is. Allow them several minutes of guided discovery with the toys. Then have them fill out the Toy Testing student page as shown:

1 Play with the toys! Then draw and label each toy below.

Toy A Drawing	Toy B Drawing

2 Give each toy a fun rating:

	Toy A			Toy B		
Fun	☹ not fun	😐 sort of fun	🙂 very fun	☹ not fun	😐 sort of fun	🙂 very fun

3 Give each toy a safety rating:

	Toy A			Toy B		
Safety	☹ not safe	😐 sort of safe	🙂 very safe	☹ not safe	😐 sort of safe	🙂 very safe

4 Which toy would you prefer to buy? Why?

Briefly compare the ratings students gave the toys. Point out that not everyone gave the toys the same ratings. Then ask

? Do you think companies use only one toy tester? (No. It is good to have more than one opinion about a toy.)

Overall Class Ratings

Discuss the idea that toy companies don't take just one person's opinion about a toy. They collect many people's opinions about a toy before making changes to the toy or before deciding to sell it in stores. Show students the Fun Ratings overhead. Point out the parts of the graph: the title, the x-axis label, the y-axis label, and the box with lines for summarizing the class ratings for Toy A and Toy B. Tell them that the graph will help them make a conclusion about the toy by showing everyone's ratings. Use a colored marker to color in the box for Toy A on the key. By a show of hands, count the number of "not fun" ratings and draw a bar

using the color for Toy A. Then count the "sort of fun" and "very fun" ratings. Next, use a different-colored marker to color in the box for Toy B on the key. By a show of hands, count the number of "not fun" ratings and draw a bar using the color for Toy B. Repeat for the other two ratings.

Have students look carefully at all of the ratings on the graph. Have them come up with an overall class fun rating for Toy A by asking

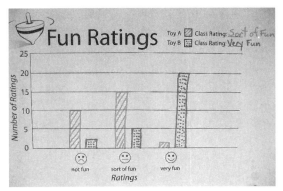

Sample fun ratings whole-class graph

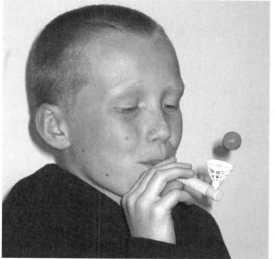

Testing a Blow Ball Pipe

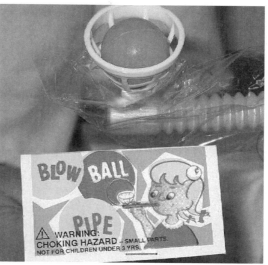

Warning label for a Blow Ball Pipe

? Which fun rating did Toy A get most often? Record that rating in the class rating box at the top of the graph. Then have students come up with an overall class fun rating for Toy B by asking

? Which fun rating did Toy B get most often?

Next, discuss what criteria students came up with to determine their safety ratings. See if students can locate a warning label on any of the toys or packaging, and discuss the possible risks of the toys. Then come up with an overall class safety rating for both toys using the Safety Rating overhead. Finally, ask students to compare the scores of both toys by comparing the class ratings. Ask

? Which toy scored higher for fun?

? Which toy scored higher for safety?

? Which toy would you prefer to buy? Why?

? How could you improve upon either of the toys?

evaluate

Toy Invention Journal and Advertising Poster

Tell students that they are going to have the op-

portunity to be toy inventors. Pass out the My Toy Invention Journal to each student. Tell them that they will be working with an adult helper at home to invent a new toy or improve a toy that they already have or know about. The journal will help them brainstorm ideas and keep track

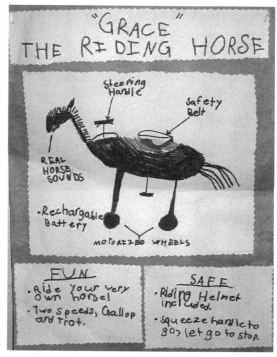

Sample toy advertisement

of their invention process. You may want to have students actually build a prototype of the new or improved toy with an adult's help. The assignment concludes with a 3-2-1 poster advertising the new or improved toy. The poster should include:

- 3 points: A labeled drawing of the new or improved toy, including a creative name for the toy.

- 2 Points: Two reasons why people should buy the toy.

- 1 Point: Directions for using the toy safely or a warning label.

- Extra Credit: A catchy slogan, a jingle, or a drawing of the toy's packaging.

You can use the rubric in the Toy Invention Journal to score completed posters and provide comments.

Inquiry Place

Have students brainstorm testable questions about toys, such as

? Which brand of toy car rolls the straightest? fastest?

? Does the size of a Frisbee affect how far it goes?

? Which brand of bubble solution makes the longest-lasting bubbles?

Then have students select a question to investigate as a class, or have groups of students vote on the question they want to investigate as a team. After they make their predictions, have them design an experiment to test their predictions. Students can present their findings at a poster session or gallery walk.

More Books to Read

Dodds, D.A. 2004. *Henry's amazing machine*. New York: Farrar, Straus, and Giroux.

Summary: From the time Henry is a baby he loves to put things together—wheels with rods, switches with levers, cranks with gears. By the age of 6, he's built an Amazing Machine that fills his entire room. By the time he's 10, the machine has taken over the entire house—and the yard. His parents are proud of Henry, but they're getting a little worried. They can't help wondering: What does it do?

Lionni, L. 1974. *Alexander and the wind-up mouse*. New York: Dragonfly Books.

Summary: Alexander the mouse finds a friend to end his loneliness—Willy the wind-up mouse. When Willy is about to be thrown away, Alexander makes a selfless decision and with the help of a magic lizard saves his friend.

McGough, R. 1997. *Until I met Dudley: How everyday things really work*. New York: Walker.

Summary: A young girl used to have fantastic ideas about how things work, but Dudley, a pencil-wielding, bespectacled dog, tells her how it really is. This lively picture book explains the inner workings of mechanical objects such as vacuum cleaners, refrigerators, dishwashers, toasters, and garbage trucks.

Taylor, B. 2003. *I wonder why zippers have teeth: And other questions about inventions*. New York: Kingfisher.

Summary: "What did people use before they had refrigerators?" and "Where do inventors get their ideas?" are some of the questions answered in this intriguing question-and-answer book about common household inventions.

Websites

Houghton Mifflin Education Place Invention Convention
www.eduplace.com/science/invention/overview.html

The History Channel History of Toys and Games
www.historychannel.com/exhibits/toys

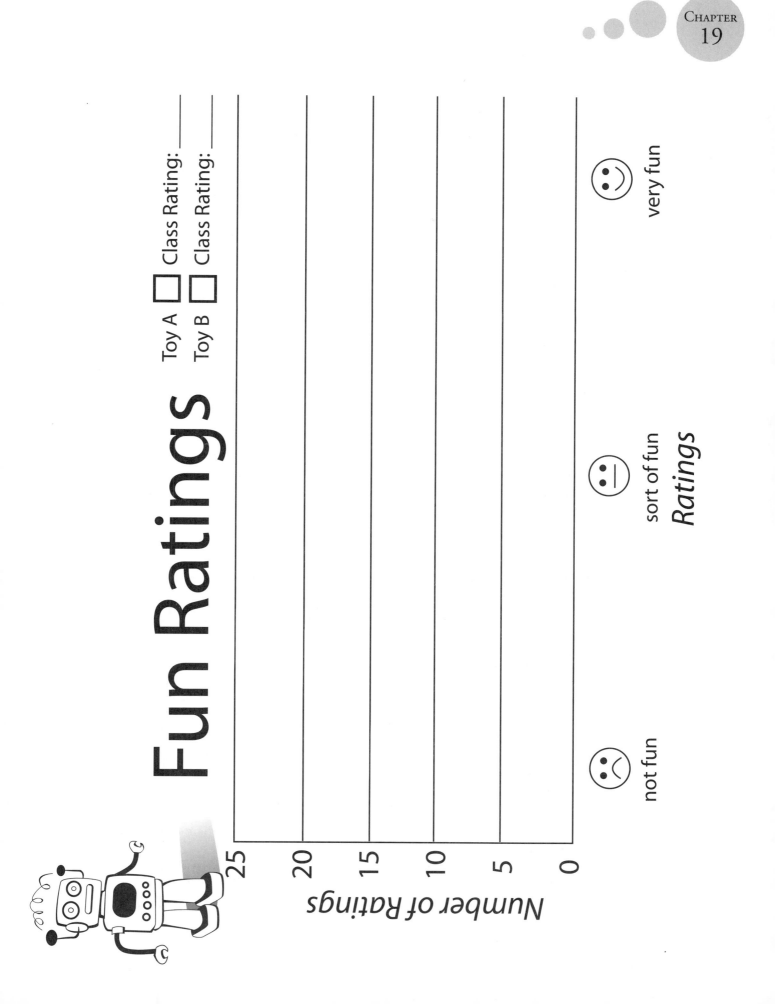

Fun Ratings

Toy A ☐ Class Rating: _____
Toy B ☐ Class Rating: _____

Number of Ratings

25
20
15
10
5
0

not fun   sort of fun   very fun

Ratings

Safety Ratings

Toy A ☐ Class Rating: _____

Toy B ☐ Class Rating: _____

Number of Ratings

25
20
15
10
5
0

🙁 not safe

😐 sort of safe

🙂 very safe

Ratings

My Improved Invention

1. Which invention would you like to improve? _____

_____

2. Draw and label your improved invention in the box below and give it a new name.

Name of My Improved Invention: _____

3. How is your improved invention more fun or useful than the original?

_____

4. What are the benefits of your improved invention?

_____

5. What are the risks of your improved invention?

_____

Name: _____

Toy Testing

You are a toy tester for Waddatoy Toys! Follow this procedure for each toy, and record your data below.

1. Play with the toys! Then draw and label each toy below.

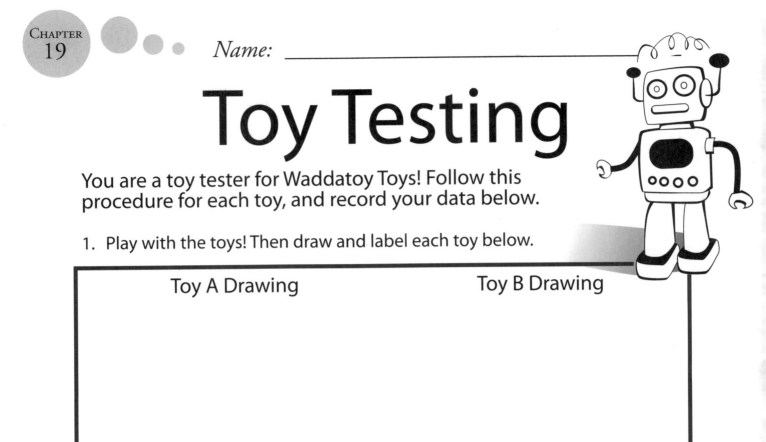

Toy A Drawing   Toy B Drawing

2. Give each toy a fun rating:

Toy A	Toy B
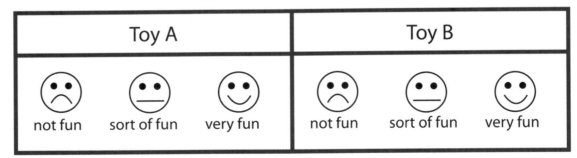	
not fun sort of fun very fun	not fun sort of fun very fun

3. Give each toy a safety rating:

Toy A	Toy B
not safe sort of safe very safe	not safe sort of safe very safe

4. Which toy would you prefer to buy? Why? _____

_____

My Toy Invention Journal

Inventor: _____

Adult Helper: _____

Brainstorming Page

Inventor

1 List some toys you like to play with:

_____   _____

_____   _____

_____   _____

_____   _____

Adult Helper

2 List some toys you liked to play with when you were a child:

_____   _____

_____   _____

_____   _____

_____   _____

Inventor and Adult Helper

3 List some toys you think could be more fun or more safe if they were improved:

_____   _____

_____   _____

_____   _____

_____   _____

Inventing Page

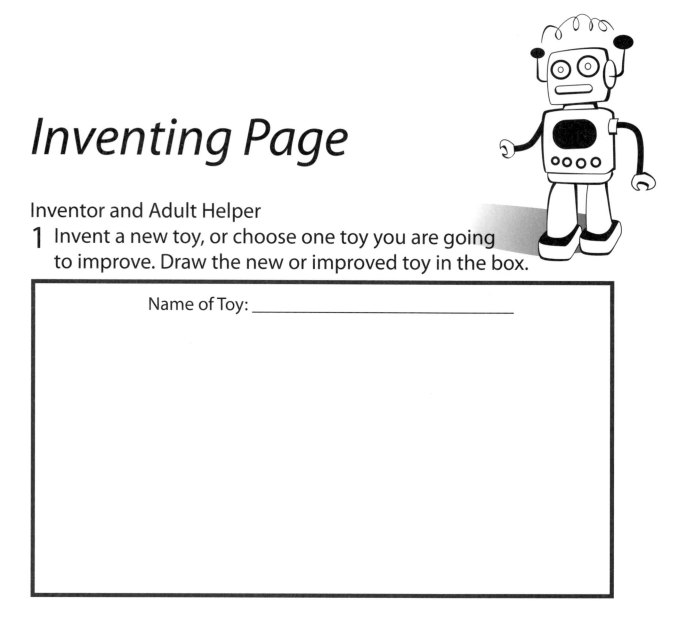

Inventor and Adult Helper

1 Invent a new toy, or choose one toy you are going to improve. Draw the new or improved toy in the box.

Name of Toy: _____

2 Fill out the T-chart to tell the benefits and risks of your new or improved toy.

Benefits	Risks

Advertising Poster Page

Make a 3-2-1 poster to advertise your new or improved toy. Your poster should include:

3   Points: A labeled drawing of the new or improved toy, including a creative name for the toy.

3   2   1   0

2   Points: Two reasons why people should buy the toy.

2   1   0

1   Point: Directions for using the toy safely or a warning label.

1   0

Extra Credit: A catchy slogan, a jingle, or a drawing of the toy's packaging.

1   0

Total Points_____/6

Comments: _____

A Sense of Wonder

Description

Learners explore the lives of two important environmentalists, John Muir and Rachel Carson. They keep nature journals to record observations and wonderings about the outdoors, as Muir and Carson did many years ago. Learners compare the characteristics of the two environmentalists and ultimately learn what it means to have a sense of wonder.

Suggested Grade Levels: 2–4

Lesson Objectives Connecting to the Standards

Content Standard C: Life Science
- Understand that humans change environments in ways that can be beneficial or detrimental for themselves and other organisms.

Content Standard G: History and Nature of Science
- Understand that many people choose science as a career and devote their entire lives to studying it.
- Understand that many people derive great pleasure from doing science.

Featured Picture Books

Title	*John Muir: America's Naturalist*	*Rachel Carson: Preserving A Sense of Wonder*	*The Important Book*
Author	Thomas Locker	Thomas Locker and Joseph Bruchac	Margaret Wise Brown
Illustrator	Thomas Locker	Thomas Locker	Leonard Weisgard
Publisher	Fulcrum Publishing	Fulcrum Publishing	HarperCollins
Year	2003	2004	1977
Genre	Narrative Information	Narrative Information	Story
Summary	A biography of John Muir, the naturalist who founded the Sierra Club and was influential in establishing the national park system.	A beautifully illustrated biography of environmentalist Rachel Carson interspersed with her own memorable quotes.	Simple poems describe everyday objects and the important thing about each of those objects.

Time Needed

This lesson will take several class periods. Suggested scheduling is as follows:

Day 1: **Engage** with *John Muir: America's Naturalist* read aloud, and **Explore/Explain** with Nature Journal Day 1.

Day 2: **Engage** with *Rachel Carson: Preserving a Sense of Wonder* read aloud, and **Explore/Explain** with Nature Journal Day 2.

Day 3: **Elaborate** with Comparing John Muir and Rachel Carson.

Day 4: **Evaluate** with *The Important Book* read aloud and The Important Thing Poster.

Materials

Hand lens (1 per student)

1 m of yarn or string (1 per pair of students)

Colored pencils, crayons, or markers

Scissors (1 per student)

Glue sticks or tape

Student Pages

My Nature Journal (Copy the cover back-to-back with pages 1 and 6; copy pages 2 and 5 back-to-back with pages 3 and 4. Fold and staple.)

Comparing John Muir and Rachel Carson

The Important Thing Poster

Background

The History and Nature of Science Standard in the National Science Education Standards focuses on the understanding of science as a human endeavor. This standard suggests that what students can learn about scientific inquiry and significant people in history will later develop into sophisticated ideas about the history and nature of science. Teachers are encouraged to use stories, videos, and other examples to introduce interesting historical examples of women and men who have made contributions to science. Another important aspect of this Standard is that many people devote their entire lives to studying science and derive great pleasure from doing science. This lesson features two major contributors to the environmental movement, John Muir and Rachel Carson. While learning about these two historical figures, students will gain an understanding of a key concept in the Life Science standard—that humans change environments in ways that can be either beneficial or detrimental for themselves and other organisms. Muir and Carson both dedicated their lives to sharing that message.

John Muir (1838–1914), one of America's most influential conservationists, was born in Scotland and immigrated to the United States in 1849. As a wilderness explorer, he is known for his extensive travels among California's Sierra Nevada Mountains and Alaska's glaciers. As a writer, he taught people about the importance of experiencing, appreciating, and protecting nature for its own sake, not just for its practical benefits to mankind. Because his writings and activism led to the creation of Yosemite, Sequoia, Mount Rainier, Petrified Forest, and Grand Canyon national parks, he is often referred to as

"The Father of our National Parks." Muir helped inspire President Theodore Roosevelt's pioneering conservation programs, including founding the first national monuments by presidential proclamation, and Yosemite National Park by congressional action. Muir and other conservationists established the Sierra Club, "to make the mountains glad." He was the club's first president and held that office until his death in 1914. The Sierra Club is now one of the most important conservation organizations in the United States.

Resource
"Who Was John Muir?"
www.sierraclub.org/john_muir_exhibit/

Rachel Carson (1907–1964), writer, scientist, and conservationist, grew up in the rural river town of Springdale, Pennsylvania. Inspired by her mother, Carson had an enduring love of nature that she first expressed as a writer and later as a student of marine biology. Her books *The Sea Around Us* (1952) and *The Edge of the Sea* (1955) made her famous as a naturalist and science writer for the public. She also wrote several articles intended to teach people about the wonder and beauty of nature, including "Help Your Child to Wonder" (1956) and "Our Ever-Changing Shore" (1957). Increasingly concerned by the use of synthetic chemical pesticides such as DDT, Carson changed the focus of her writing to caution the public about the long-term effects of misusing pesticides. Her groundbreaking book *Silent Spring* (1962) is often credited with having launched the global environmental movement. In the book, Carson challenged the practices of agricultural scientists and the government and called for a change in the way humans viewed nature. Although Carson was attacked by the chemical industry and some in government as an alarmist, she courageously spoke out with the message that humans are a vulnerable part of the natural world, subject to the same damage as the rest of the ecosystem. Carson testified before Congress in 1963, calling for new policies to protect human health and the environment. She never lived to see the banning of DDT. In 1964, Rachel Carson died after a long battle against breast cancer. Her writings continue to inspire new generations to protect the living world and all its creatures. In 1980, she was posthumously awarded the highest civilian honor in the United States, the Presidential Medal of Freedom.

Resource
A biographical entry of Linda Lear, 1998, author of *Rachel Carson: Witness for Nature* (1997) at *www.rachelcarson*.org

engage

John Muir: America's Naturalist Read Aloud

Inferring

Show students the cover of *John Muir: America's Naturalist*. Ask

? What do you think this book might be about?

Read the title and the author and ask

? What do you think a *naturalist* is? Are there clues from the word and the cover of the book that might help us figure out what a naturalist is? (Naturalist contains the word *nature*. Trees, a person, and a bear are pictured on the cover, so a naturalist might be a person who studies nature.)

Making Connections: Text to Self

Read the book aloud. As you read, stop at the pages suggested below and ask

? (p. 6) What does nature mean to you? (Answers might include: something that is not man-made or something natural or wild.)

? (p. 12) Have you ever kept a nature journal?

? (p. 14) Have you ever seen something so beautiful in nature that it took your breath away? Describe it to a partner.

? (p.18) Do you like to make drawings of animals or trees? What do you like to draw?

? (p. 30) Have you ever wandered in a wild place?

explore/explain

Nature Journal Day 1

After reading, pass out a My Nature Journal to each student. Have students write their names on the front cover and then open to pages 1 and 2. Tell them that they are going outdoors to a "wild" place and, like John Muir did, record their observations in pictures and words.

SAFETY Check your district policy on taking students outside during the school day or away from the school grounds before doing the following activity.

Read the following quote from John Muir: "A thousand … wonders are calling. Look up and down and round about you." Explain to students that they will be taking a nature walk to look up and down and around to observe some wonders of nature.

Next, take students outside on a short walk with their journals and pencils to a "wild" place (this could be a grassy area, a grove of trees or even a patch of sidewalk or blacktop that has grass or weeds growing near it). Point out several natural things that students could choose to observe, such as the sky, trees, bushes, ants, and weeds. Have students look all around for these natural wonders. Ask students to be silent during the walk so that they can notice

Recording observations in journals

the sounds of nature around them. Model your observations and questions to the students:

Observations

I hear birds chirping in the trees.

Some of the leaves on the trees are green, and some are turning yellow.

Questions

? Where do the birds go at night?

? How old are these trees?

Have students choose a place to observe, sit down where they can get a good view, record as much detail as possible through drawing and recording observations, and then write some of their questions about the place. When you return to the classroom, have students share pages 1 and 2 of their journals with a partner. Students will use the journals again in the next activity.

engage

Rachel Carson Read Aloud

Inferring

Show students the cover of *Rachel Carson: Preserving a Sense of Wonder*. Ask

? What do you think this book might be about?

Read the title and the author and ask

? What do you think the phrase "a sense of wonder" means? Are there clues from the cover of the book that might help us figure out what it means? (A woman is pictured looking at a starfish in her hand; maybe she is wondering about it or is fascinated by it.)

Making Connections: Text to Self

? (p.12) Why do you think Rachel chose to study living things? (She loved nature.) Have you thought about what you would like to study when you are older?

? (p.18) What changes have you noticed in nature this season? (Answers might include

Observing nature up close

that, depending on the season, students may have noticed changes in animal behavior and appearances of plants.)

? (p.24) Have you ever seen a wild place that was harmed by human actions? Describe it to a partner. (Students might describe places where they have seen litter, deforestation, or pollution.)

explore/explain

Nature Journal Day 2

After reading, pass out the students' nature journals and have them turn to page 3. Tell them that

SAFETY
Check your district policy on taking students outside during the school day or away from the school grounds before doing the following activity.

they are going to go outside to observe nature again but this time on a smaller scale. Tell them that there are some very small wild things outdoors that they might not have noticed on their first nature walk. Give each student a hand lens and each pair of students 1 m of string. Tell them that, when they go outside, they will be looking down on the ground for some natural things to observe. With their partner, they will shape their piece of string into a circle on the ground. They will use their hand lenses to carefully observe the living and nonliving things inside the string and draw a picture of what they see. Then they will record their observations and questions on page 4.

Take students outdoors, and model for them how to place their strings in a circle on the ground and how to use their hand lenses to get a closer look at things. Have students choose an area, shape their string into a circle, and work in silence. When you return to the classroom, have students share pages 3 and 4 of their journals with a partner.

Comparing John Muir and Rachel Carson

elaborate

Comparing John Muir and Rachel Carson

Synthesizing

In pairs, have students read the quotes on pages 5 and 6 of their journals.

John Muir

"All that the sun shines on is beautiful, so long as it is wild."

"When we are with Nature, we are awake, and we discover many interesting things and reach many a mark we are not aiming at."

Rachel Carson

"Every mystery solved brings us to the threshold of a greater one."

"The more clearly we can focus our attention on the wonders and realities of the world about us,

the less taste we shall have for destruction."

Have students work with their partners to restate the quotes in their own words, and then try to figure out their deeper meaning. Discuss student thinking as a class. Then ask

? Why do you think John Muir and Rachel Carson dedicated their lives to preserving nature? (They loved nature and were curious about natural things. They were also amazed at the beauty of the natural world. They found happiness being in natural places.)

Revisit the book *Rachel Carson: Preserving a Sense of Wonder,* and read the entire title aloud. Ask

? What do you think it means to have "a sense of wonder"? (to be curious about or amazed by something).

Explain that the word *wonder* can mean exciting amazed admiration. Model statements that express *exciting amazed admiration*, such as

- It is exciting to think that this rock may have been on Earth for millions of years!

- I am amazed at how blue the sky looks today.

- I admire how these tiny ants can lift things that are many times heavier than their own bodies.

Then ask

? How did Rachel Carson demonstrate a sense of wonder? (Answers might include: She marveled at living things she saw in a microscope, she looked at the ocean with amazement, and she wrote about nature's wonders.)

? How did John Muir demonstrate a sense of wonder? (Answers might include: The beauty of Yosemite took his breath away, he studied the ways of the animals, and he was excited about the wildness of storms and winter.)

Making Connections: Text to Text

Hold up both books and ask

? How are these books similar? (Answers might include: similar illustrations and about people

who loved nature. Students may notice that both books are illustrated with paintings by Thomas Locker.)

Ask

? What do John Muir and Rachel Carson have in common? (Answers might include: They both loved nature, were writers, and loved the outdoors.)

Tell students that one tool they can use to compare how things are alike and different is a Venn diagram. Pass out the Comparing John Muir and Rachel Carson student page. Model how one student can read a John Muir fact, while another looks to see if there is a Rachel Carson fact that is similar. If similar, they can place both facts in the middle where the circles intersect. If the John Muir fact doesn't match up with a similar Rachel Carson fact, they can place it in the John Muir circle. Next, have students cut out each fact and begin comparing them. Before students glue or tape the facts on their papers, go over the correct placement of the statements as a class.

Comparing John Muir and Rachel Carson

John Muir	Both	Rachel Carson
John was born in Scotland in 1838 and moved to Wisconsin when he was 11.	John lived on a farm when he was young.	Rachel was born in Pennsylvania in 1907.
John loved exploring the mountains.	Rachel lived on a farm when she was young.	Rachel loved studying the ocean.
John was supported by many people in his fight to create national parks.	John wrote his observations in a journal. Rachel wrote her observations in a journal.	Rachel was criticized by many people for writing *Silent Spring*, about the dangers of spraying pesticides (poisons).
John is known as the father of our National Parks.	John devoted his life to changing the way people thought about nature. Rachel devoted her life to changing the way people thought about nature.	Rachel is known as the founder of the Environmental Movement.

The Important Book *read aloud*

evaluate

The Important Book Read Aloud and The Important Thing

Discuss some of the facts students have learned about John Muir and Rachel Carson. Tell students that, as time passes, it might be difficult to remember all of these facts about the two naturalists. Therefore it is helpful to think about the most important thing about each of these people rather than all the details.

Determining Importance

Show students the cover of *The Important Book* and tell them that this book might help us think about how to determine the *most* important thing about John Muir and Rachel Carson. Read *The Important Book* aloud to students and ask them to listen for "the important thing" about each object.

Discuss the idea that John Muir and Rachel Carson both had a positive impact on the way people thought about nature. Both naturalists dedicated their lives to preserving nature. Ask

? In what ways can humans harm nature?

? In what ways can humans help nature?

? Why is it important for humans to take care of nature? (Answers might include: Humans depend on nature for food, clean air, and clean water.)

Pass out The Important Thing student page to each student. Tell them that they are going to make a poster similar to the pages in The Important Book about either Rachel Carson or John Muir. Have students refer to the Venn diagram they created in the previous lesson to help them determine the important thing about the naturalist they choose. They must also include an illustration of the important thing about that person.

Scoring Rubric for The Important Thing Student Page

4 Point Response	The student includes all five required elements: a drawing of the person, three facts about the person, and an important characteristic or achievement of the person.
3 Point Response	The student omits one required element.
2 Point Response	The student omits two required elements.
1 Point Response	The student omits three or four required elements.
0 Point Response	The student omits all five required elements.

Inquiry Place

Have students brainstorm testable questions about the natural world, such as

- What changes can we see in our "wild place" from day to day or season to season?

- What will we find more of in our "wild place": plants or animals? insects or arachnids? birds or mammals? Or something else?

- What evidence can we see of human impact on our "wild place"? Is it positive, negative, or both? How can we protect or care for our "wild place"?

Then have students select a question to investigate as a class, or have groups of students vote on the question they want to investigate as a team. After they make predictions, have them design an experiment to test their predictions. Students can present their findings at a poster session or gallery walk.

More Books to Read

Atkins, J. 2000. *Girls who looked under rocks.* Nevada City, CA: Dawn Publications.

Summary: Six girls, from the 17th to the 20th century, didn't run from spiders or snakes but crouched down to take a closer look. They became pioneering naturalists, passionate scientists, and energetic writers or artists.

Ehrlich, A. 2003. *Rachel: The story of Rachel Carson.* New York: Harcourt.

Summary: Rachel Carson was always curious about the world around her. As a girl she loved being outside, always exploring and wanting to know more about the universe. As an adult Rachel wrote books about what she loved—including *Silent Spring*, a book that brought to the world an awareness of the dangers of pesticides.

Fredericks, A. D. 2001. *Under one rock: Bugs, slugs, and other ughs.* Nevada City, CA: Dawn Publications.

Summary: Explore the fascinating community of creatures that live under one rock. Rhythmic verse and colorful close-up illustrations draw the reader into the incredible world of bugs, spiders, and creepy-crawlies.

George, L. B. 1995. *In the snow: Who's been here?* New York: Mulberry Books.

Summary: Two children on their way to go sledding see evidence of a variety of animal life. The reader must infer from the evidence what animals had been in each location. Each time, the answers are revealed on the next page.

George, L. B. 1995. *In the woods: Who's been here?* New York: Mulberry Books.

Summary: A boy and girl in the autumn woods find an empty nest, a cocoon, gnawed bark, and other signs of unseen animals and their activities. The reader must infer from the evidence what animals had been in each location. Each time, the answers are revealed on the next page.

Pratt-Serafini, K. J. 2001. *Salamander rain: A lake and pond journal.* Nevada City, CA: Dawn Publications.

Summary: Journal notes, maps, illustrations, newspaper clippings, and text provide a rich description of seasonal changes that occur in a temperate pond community. The book acts as an invitation for readers to explore and learn more about wetlands and provides a wonderful model for student journaling.

Rubay, D. 1998. *Stickeen: John Muir and the brave little dog.* Nevada City, CA: Dawn Publications.

Summary: This is a captivating introduction to the life and values of famous environmentalist John Muir, adapted from Muir's original account. The beautiful illustrations by Christopher Canyon are integral in telling the story about an incident that occurred when Muir was exploring Alaska. Trapped on a glacier, he made his way across a dangerous ice bridge, followed by an independent-minded dog named Stickeen.

Websites

Sierra Club
www.sierraclub.org

A Website Devoted to the Life and Legacy of Rachel Carson
www.rachelcarson.org

Dawn Publications: Sharing Nature with Children
www.dawnpub.com

My Nature Journal

Naturalist: _____

Rachel Carson Quotes

"Every mystery solved brings us to the threshold of a greater one."

"The more clearly we can focus our attention on the wonders and realities of the world about us, the less taste we shall have for destruction."

6

Look Around

Date: _____ Time: _____

1

Observations

_____
_____
_____
_____

Questions

_____
_____
_____
_____
_____

John Muir Quotes

"All that the sun shines on is beautiful, so long as it is wild."

"When we are with Nature, we are awake, and we discover many interesting things and reach many a mark we are not aiming at."

Observations

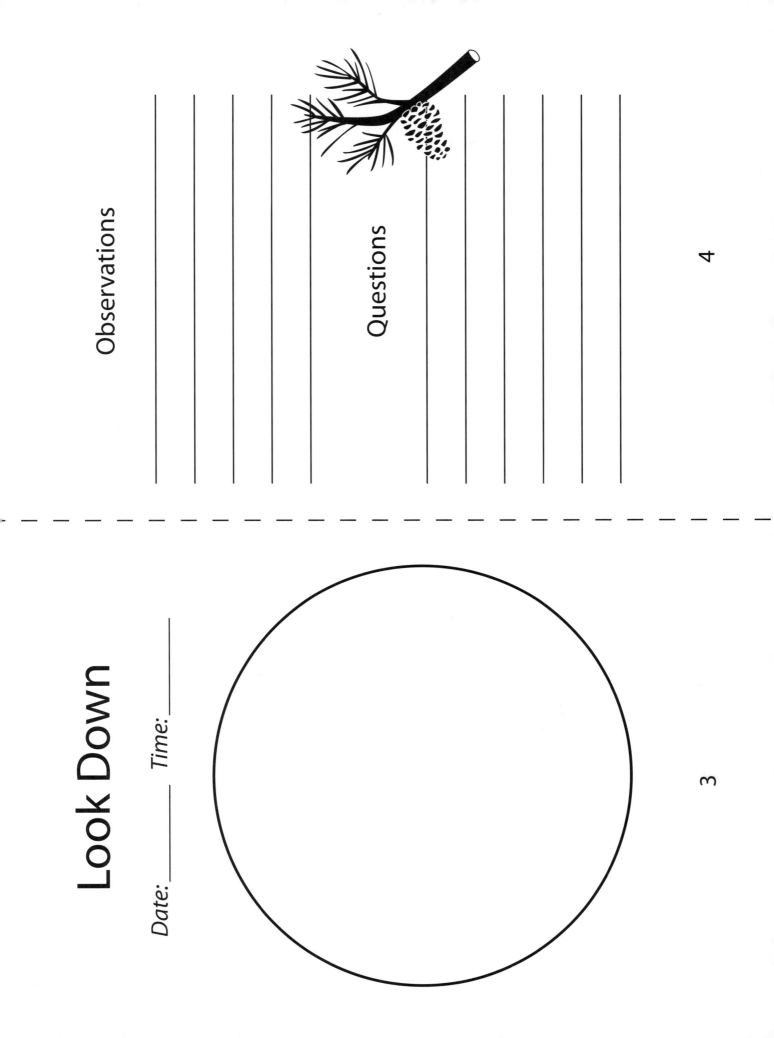

Questions

Look Down

Date: _____ *Time:* _____

Name: _____

John Muir and
Rachel Carson

Directions: Cut out these facts about famous naturalists John Muir and Rachel Carson. Then place each one on the Venn Diagram to show how they are alike and different.

John wrote his observations in a journal.	Rachel was born in Pennsylvania in 1907.
John devoted his life to changing the way people thought about nature.	Rachel loved studying the ocean.
John is known as the father of our national parks.	Rachel devoted her life to changing the way people thought about nature.
John was supported by many people in his fight to create national parks.	Rachel was criticized by many people for writing *Silent Spring*, a book about the dangers of spraying pesticides (poisons).
John lived on a farm when he was young.	Rachel lived on a farm when she was young.
John loved exploring the mountains.	Rachel wrote her observations in a journal.
John was born in Scotland in 1838 and moved to Wisconsin when he was 11.	Rachel is known as the founder of the Environmental Movement.

Comparing

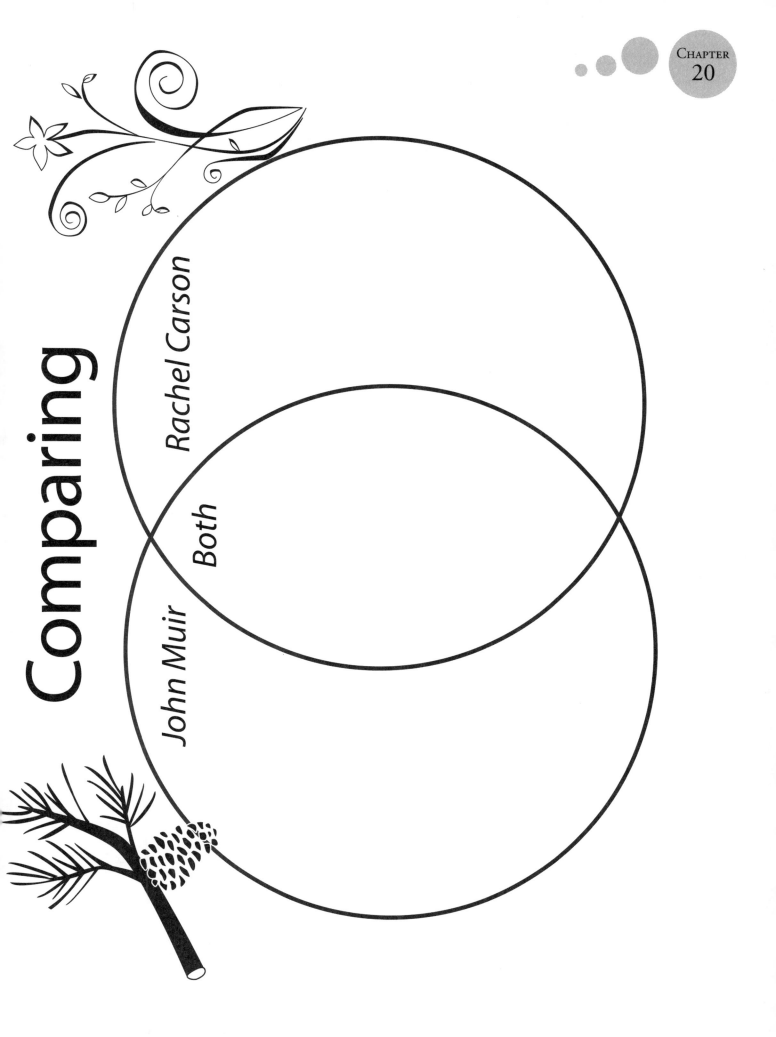

Rachel Carson

Both

John Muir

Name: _____

The Important
Thing About

_____

[ ]

The important thing about _____

is _____ .

Some facts about _____ are

1. _____

_____

2. _____

_____

3. _____

But the important thing about _____ is

_____ .

Glossary

Anticipation guides—Sets of questions that serve as a pre- or post-reading activity for a text, anticipation guides can activate and assess prior knowledge, determine misconceptions, focus thinking on the reading, and motivate reluctant readers by stimulating interest in the topic.

Assessment—"Assessment, broadly defined, means information gathering. Grading (or evaluating) students is certainly one type of assessment. Tests, portfolios, and lab practicals are all assessment devices. However, teachers assess students in other ways. When teachers check for understanding, to determine whether or not to continue teaching about a particular idea and where to go next with instruction, they are also assessing their students. Ungraded pretests and self-tests likewise represent assessment. Any information that helps the teacher make instructional decisions is assessment.

"Assessment is valuable to students as well as teachers (not to mention parents and other education stakeholders) because it helps students figure out what they do and don't understand and where they need to place their efforts to maximize learning. Assessment is also used to sort or rank students, letting them know how their performance compares to others, both for placement purposes and as a way to ensure minimum competencies in those who have passed particular tests." (Colburn 2003, p. 37)

Chunking—Chunking, just like it sounds, is dividing the text into manageable sections and reading only a section at any one time.

Cloze paragraph—An activity to help readers infer the meaning of unfamiliar words. Key words are deleted from a passage, and students fill in the blanks with words that make sense and sound right.

Constructivism—"Constructivism has multiple meanings, and it's important that when people discuss the concept they be sure they're talking about the same thing! Much of the confusion stems from the fact that constructivism refers to both an explanation (theory) about how people learn and a philosophical position related to the nature of learning (see Matthews 1994, pp. 137–39). Increasingly, people are also using the term to refer to teaching techniques designed to build on what students already know, for example, open-ended, hands-on inquiry (Brooks and Brooks 1993).

"I'd like to focus on constructivism as an explanation about learning; that's probably what is most relevant to readers. In this context, *constructivism* refers to the concept that learners always bring with them to the classroom (or any other place where learning takes place) ideas about how the world works—including ideas related to whatever may be in today's lesson. Most of the time learners are unaware they even have these ideas! The ideas come from life experiences combined with what people have learned elsewhere.

"According to constructivist learning theory, learners test new ideas against that which they already believe to be true. If the new ideas seem to fit in with their pictures of the world, they have little difficulty learning the ideas. There's no guarantee, though, that they will fit the ideas into their pictures of how the world works with the kind of meaning the teacher intends. . . .

"On the other hand, if the new ideas don't seem to fit the learner's picture of reality then they won't seem to make sense. Learners may dismiss them, learn them well enough to please the teacher (but never fully accept the ideas), or eventually accommodate the new ideas and change the way they understand the world. As you might guess, this third outcome is most difficult to achieve, although it's what teachers most often desire in students.

"Seen this way, teaching is a process of trying to get people to change their minds—difficult enough as is, but made even more difficult by the fact that learners may not even know they hold an opinion about the idea in question! People who study learning and cognition often contrast constructivism with the more classical idea that students in our classes are "blank slates" who know nothing about the topics they are being taught. From this perspective, the teacher "transmits" new information to students, who mentally store it away. In contrast, constructivist learning theory says that students are not blank slates; learning is sometimes a process whereby new ideas help students to 'rewrite' the misconceptions already on their slates." (Colburn 2003, pp. 58–59)

Determining importance—One of Harvey and Goudvis' six key reading strategies, determining importance involves identifying essential information by distinguishing it from nonessential details. (See also Inferring, Making connections, Questioning, Synthesizing, and Visualizing.)

Dual-purpose books—Intended to serve two purposes, present a story and provide facts, dual-purpose books employ a format that allows readers to use the book as a storybook or as a non-narrative information book. Sometimes information can be found in the running text but more frequently appears in insets and diagrams. Readers can enter on any page to access specific facts or read the book through as a story.

Elaborate—See 5E model of instruction.

Engage—See 5E model of instruction.

Evaluate—See 5E model of instruction.

Explain—See 5E model of instruction.

Explore—See 5E model of instruction.

Features of nonfiction—Many nonfiction books include a table of contents, index, glossary, bold-print words, picture captions, diagrams, and charts that provide valuable information. Modeling how to interpret the information is important because children often skip over these features.

5E model of instruction—"The 5E model of instruction is a variation on the learning cycle model, pioneered by the Biological Sciences Curriculum Study (BSCS 1993). The five Es of the model are *engage, explore, explain, elaborate,* and *evaluate. Engage* refers to beginning instruction with something that both catches students' attention and helps them relate what is to come with what they already know. *Explore* is virtually identical with the exploration phase of the learning cycle, as *explain* is the concept- or term-introduction phase and *elaborate* is the application phase. *Evaluation* is both formative and summative since it helps determine whether instruction should continue or whether students need more time and teaching to learn the unit's key points." (Colburn 2003, p. 23)

Genre—Picture books are a genre in themselves, but in this text, genre refers to types of picture books: storybooks, non-narrative information books, narrative information books, and dual-purpose books.

Guided inquiry activity—"In a guided inquiry activity, the teacher gives students only the problem to investigate (and the materials to use for the investigation). Students must figure out how to answer the investigation's question and then generalize from the data collected." (Colburn 2003, pp. 20–21)

Inferring—One of Harvey and Goudvis' six key reading strategies, inferring involves merging clues from the reading with prior knowledge to draw conclusions and interpret text. (See also Determining importance, Making connections, Questioning, Synthesizing, and Visualizing.)

Inquiry—"Historically, discussions of inquiry generally have fallen within two broad classes. Sometimes people talk about inquiry as describing what scientists do and sometimes as a teaching and learning process. Authors of the *National Science Education Standards* (NRC 1996) seemed to recognize this dichotomy:

Scientific inquiry refers to the diverse ways in which scientists study the natural world and propose explanations based on the evidence derived from their work. Inquiry also refers to the activities of students in which they develop knowledge and understanding of scientific ideas, as well as an understanding of how scientists study the natural world. [emphasis added] (23)

"To make this distinction less confusing, people also sometimes use the phrase 'inquiry-based instruction.' This term refers to the creation of a classroom where students are engaged in (essentially) open-ended, student-centered, hands-on activities. This means that students must make at least some decisions about what they are doing and what their work means—thinking along the way.

"While most people in the science education community would probably think of inquiry as hands-on, it's also true that many educators would 'count' as inquiry any activity where students are analyzing real-life data—even if the information were simply given to students on paper, without any hands-on activity on their part.

"As readers can begin to see, inquiry and inquiry-based instruction represent ideas with broad definitions and occasional disagreements about their meaning. Two people advocating inquiry-based instruction may not be advocating for the same methods! Some define 'inquiry' (instruction) in terms of open-ended, hands-on

instruction; others define the term in terms of formally teaching students inquiry skills (trying to teach students how to observe or make hypotheses, for example); and some define inquiry so broadly as to represent any hands-on activity." (Colburn 2003, pp. 19–20)

Learning cycle—"Different versions of the learning cycle exist today. However, the general pattern is to begin instruction with students engaged in an activity designed to provide experience with a new idea. The idea behind this exploratory phase of the cycle is that learning of new ideas is maximized when students have had relevant, concrete experience with an idea before being formally introduced to it (Barman and Kotar 1989).

"This exploratory phase is ideally followed by a concept- or term-introduction phase. That phase generally begins with class discussion about student findings and thoughts following the previous part of the cycle. Sometimes the teacher can then go on to simply provide names for ideas that students previously discovered or experienced.

"Finally, students expand on the idea in an application phase of instruction in which they use the new idea(s) in a different context. Using a new idea in a new context is an important part of maximizing learning. In addition, some students don't begin to truly understand an idea until they've had the time to work with it for a while, in different ways. The learning cycle model provides these students with time and opportunities that help them learn.

"Ideally, the application phase of the cycle also introduces students to a new idea. In this sense, the application phase of one learning cycle is also the exploratory phase of another learning cycle—hence the 'cycle' part of 'learning cycle.' (Notice that the previous sentence began with the word 'ideally'; sometimes it's difficult for an application phase activity to also encourage students to explore other ideas.)" (Colburn 2003, p. 22)

Making connections—One of Harvey and Goudvis' six key reading strategies, making connections involves having students access their prior knowledge and experience to make meaningful connections to the text when reading. Three types of connections are Text to Self, Text to Text, and Text to World. (See also Determining importance, Inferring, Questioning, Synthesizing, and Visualizing.)

Misconceptions—"[L]earners always bring preconceived ideas with them to the classroom about how the world works. Misconceptions, in the field of science education, are preconceived ideas that differ from those currently accepted by the scientific community. Educators use a variety of phrases synonymously with 'misconceptions,' including 'naive conceptions,' 'prior conceptions,' 'alternate conceptions,' and 'preconceptions.' Many people have interviewed students to discover commonly held scientific ideas (Driver, Guesne, and Tiberghien 1985; Osborne and Freyberg 1985)." (Colburn 2003, p. 59)

Most valuable point (MVP)—An activity in which students are asked to determine the most valuable point after reading a passage. The purpose of this tool is to help readers distinguish between unimportant and important information in order to identify key ideas as they read.

Narrative information books—Narrative information books communicate a sequence of factual events over time and sometimes recount the events of a specific case to generalize to all cases. Teachers should establish a purpose for reading so students focus on the science content rather than the storyline. Teachers may want to read the book through one time for the aesthetic components and a second time for specific science content.

National Science Education Standards—"The National Science Education Standards were published in 1996, after a lengthy commentary period from many interested citizens and groups . . .

"The Standards were designed to be achievable by all students, no matter their background or characteristics. . . .

"Beside standards for science content and for science teaching, the *National Science Education Standards* includes standards for professional development for science teachers, science education programs, and even science education systems. Finally, the document also addresses what some consider the bottom line for educational reform—standards for assessment in science education.

"Although the information in the *National Science Education Standards* is often written in a rather general manner, the resulting document provides a far-reaching and generally agreed upon comprehensive starting place for people interested in changing the U.S. science educational system." (Colburn 2003, pp. 81–82)

Non-narrative information books—Factual texts that introduce a topic, describe the attributes of the topic, or describe typical events that occur. The focus is on the subject matter, not specific characters. The vocabulary is typically technical, and readers can enter the text at any point in the book.

Open inquiry activity—"Open inquiry, in many ways, is analogous to doing science. Problem-based learning and science fair activities are often open inquiry experiences for students. Basically, in an open inquiry activity students must figure out pretty much everything. They determine questions to investigate, procedures to address their questions, data to generate, and what the data mean." (Colburn 2003, p. 21)

O-W-L chart—An O-W-L chart ("Observations, Wonderings, Learnings") is one of several organizers that can help learners activate prior knowledge, organize their thinking, understand the essential characteristics of concepts, and see relationships among concepts. It can be used for prereading, for assessment, or for summarizing or reviewing material.

Pairs read—In a pairs read, one learner reads aloud, while the other listens and then summarizes the main idea. Benefits include increased reader involvement, attention, and collaboration and students who become more independent learners.

Picture walk—An activity in which the teacher shows students the cover of a book and browses through the pages, in order, without reading the text. The purpose of this tool is to establish interest in the story and expectations about what is to come.

Questioning—One of Harvey and Goudvis' six key reading strategies, questioning involves readers asking themselves questions before, during, and after reading. This allows readers to construct meaning, find answers, solve problems, and eliminate confusion as they read. (See also Determining importance, Inferring, Making connections, Synthesizing, and Visualizing.)

Questioning the author (QtA)—An interactive strategy that encourages students to question the ideas presented in the text while they are reading, making them critical thinkers, not just passive readers.

Reading aloud—Being read to builds knowledge for success in reading and increases interest in reading and literature and in overall academic achievement. See Chapter 2 for more on reading aloud, including 10 tips on how to do it.

Reading comprehension strategies—The six key reading comprehension strategies featured in *Strategies That Work* (Harvey and Goudvis 2000) are *making connections, questioning, visualizing, inferring, determining importance,* and *synthesizing*. See Chapter 2 for fuller explanations.

Rereading—Nonfiction text is often full of unfamiliar ideas and difficult vocabulary. Rereading content for clarification is an essential skill of proficient readers, and you should model this frequently. Rereading for a different purpose can aid comprehension. For example, a teacher might read aloud for enjoyment and then revisit the text to focus on science content.

Sketch to stretch—Learners pause briefly to reflect on the text and do a comprehension self-assessment by drawing on paper the images they visualize in their heads during reading. Teachers should have students use pencils so they understand the focus should be on collecting their thoughts rather than creating a piece of art. You may want to use a timer.

Stop and jot—Learners stop and think about the reading and then jot down a thought. If they use sticky notes, the notes can be added to a whole-class chart to connect past and future learning.

Storybooks—Storybooks center on specific characters who work to resolve a conflict or problem. The major purpose of stories is to entertain. The vocabulary is typically commonsense, everyday language. A storybook can spark interest in a science topic and move students toward informational texts to answer questions inspired by the story.

Structured inquiry activity—"In a structured inquiry activity, the teacher gives students a (usually) hands-on problem they are to investigate, and the methods and materials to use for the investigation, but not expected outcomes. Students are to discover a relationship and generalize from data collected.

"The main difference between a structured inquiry activity and verification lab (or 'cookbook activity') lies in what students do with the data they generate. In structured inquiry activities, students are largely responsible for figuring out what the data might mean—that is, they analyze and interpret the data. Students may ultimately interpret the data differently; different students may come to somewhat different conclusions. In a verification lab, on the other hand, all students are expected to arrive at the same conclusion—there's a definite right answer that students are supposed to be finding during the lab activity." (Colburn 2003, p. 20)

Synthesizing—One of Harvey and Goudvis' six key reading strategies, synthesizing involves combining information gained through reading with prior knowledge and experience to form new ideas. (See also Determining importance, Inferring, Making connections, Questioning, and Visualizing.)

T-chart—A T-chart is one of several organizers that can help learners activate prior knowledge, organize their thinking, understand the essential characteristics of concepts, and see relationships among concepts. It can be used for prereading, for assessment, or for summarizing or reviewing material.

Turn and talk—Learners pair up with a partner to share ideas, explain concepts in their own words, or tell about a connection they have to the book. This method allows each child to be involved as either a talker or a listener.

Using features of nonfiction—An activity in which the teacher models how common features of nonfiction can be used to help the reader. These features include table of contents, index, glossary, bold-print words, picture captions, diagrams, and charts.

Venn Diagram—A graphic organizer, made of two or more intersecting circles, that is used to compare two or more items, such as books, people, animals, or events.

Visualizing—One of Harvey and Goudvis' six key reading strategies, visualizing involves creating mental images while reading or listening to text. This strategy can help engage learners and stimulate their interest in the reading. (See also Determining importance, Inferring, Making connections, Questioning, and Synthesizing.)

References

Barman, C. R., and M. Kotar. 1989. The learning cycle. *Science and Children* (April): 30–32.

Biological Sciences Curriculum Study (BSCS). 1993. *Developing biological literacy.* Dubuque, IA: Kendall/Hunt.

Brooks, J. G., and M. G. Brooks. 1993. *In search of understanding: The case for constructivist classrooms.* Alexandria, VA: Association for Supervision and Curriculum Development.

Colburn, A. 2003. *The lingo of learning: 88 education terms every science teacher should know.* Arlington, VA: NSTA Press.

Driver, R., E. Guesne, and A. Tiberghien. 1985. *Children's ideas in science.* Buckingham, England: Open University Press.

Harvey, S., and A. Goudvis. 2000. *Strategies that work: Teaching comprehension to enhance understanding.* York, ME: Stenhouse Publishers.

Matthews, M. R. 1994. *Science teaching: The role of history and philosophy of science.* New York: Routledge.

National Research Council (NRC). 1996. *National science education standards.* Washington, DC: National Academy Press. Available online at *http://books. nap.edu/html/nsts/html/index.html.*

Osborne, R., and P. Freyberg. 1985. *Learning in science.* Portsmouth, NH: Heinemann.

Index